FOOD CHEMISTRY

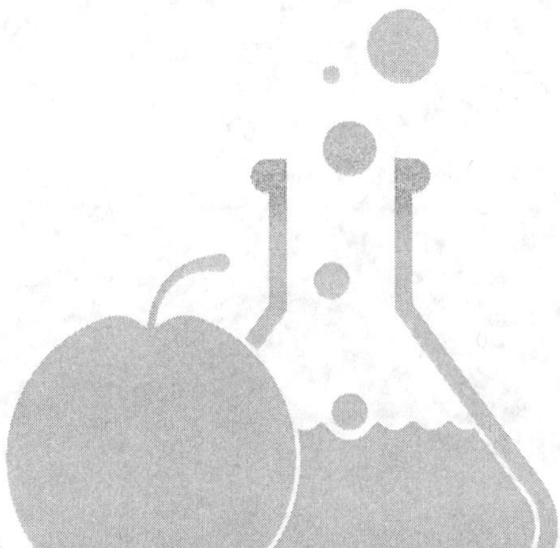

The Author

Shalini Saxena is Head of the Department of Home Science at Dyanand University, Karnataka. She completed her M. Sc. in Food Technology from Indian Institute of Technology, Chennai in 1981 and completed her Ph. D. from Annamalai University in 1988. She has been teaching the subject since 1980. Her areas of interests are in Thermal Processing (Conventional Canning, Agitation Processing, Micro-Wave Processing), Biochemical changes in Food, Water activity in Food, Calculation of the energy content of foods, Methods of Food preservation, Chemistry of Food Irradiation.

FOOD CHEMISTRY

Shalini Saxena

RAJAT PUBLICATIONS

New Delhi - 110002

Food Chemistry
by Shalini Saxena

This edition is published by Rajat Publications
4740/23, Ansari Road Daryaganj,
New Delhi - 110002

ISBN: 978-81-7880-422-4

Edition 2026

Printed and bound in India

RAJAT
PUBLICATIONS

4740/23, Ansari Road, Daryaganj,
New Delhi- 110002 (India)
Tel:- +91-11-23267924
Mob:- +91-9811484973, +91-9811105389
E-mail: rajat_publication@yahoo.com

Preface

Food Chemistry is the study of the Chemistry of Foods, their deterioration, and the principles underlying the improvement of foods for the consuming public. It is the application of chemistry to the Development, Processing, Packaging, Preservation, Storage, and distribution of Foods and Beverages for the purposes of obtaining a safe, Economical, and Aesthetically pleasing supply of food for people worldwide. Food Science involves the application of Chemistry, Biology, Physics, Biochemistry, Microbiology, Nutrition, and Engineering to the development and distribution of food, the major portion of a food science curriculum is Chemistry.

This book provides a wide range of information about the Chemical and Biochemical Composition of Foods, which is fundamental to the study of their properties and processing applications. The contents of the book will be of immense value to Students, Researchers, Food Scientists and Technologists.

-Author

Contents

CHAPTER 1

Introduction

Food chemistry is the study of chemical processes and interactions of all biological and non-biological components of foods. The biological substances include such items as meat, poultry, lettuce, beer, and milk as examples. It is similar to biochemistry in its main components such as carbohydrates, lipids, and protein, but it also includes areas such as water, vitamins, minerals, enzymes, food additives, flavors, and colors. This discipline also encompasses how products change under certain food processing techniques and ways either to enhance or to prevent them from happening. An example of enhancing a process would be to encourage fermentation of dairy products with lactic acid; an example of a preventing process would be stopping the Maillard reaction on the surface of freshly cut Red Delicious apples whether by hand or mechanical methods.

HISTORY OF FOOD CHEMISTRY

Food chemistry's history dates back as far as the late 1700s when many famous chemists were involved in discovering chemicals important in foods, including Carl Wilhelm Scheele, and Sir HumpHry Davy published the first ever book on agricultural and food chemistry in 1813 titled Elements of Agricultural Chemistry, in a Course of Lectures for the Board of Agriculture in the United Kingdom which would serve as a foundation for the profession worldwide, going into a fifth edition.

In 1874 the Society of Public Analysts was formed, with the aim of applying analytical methods to the benefit *of* the public. It's early experiments were based on bread, milk and wine. It was also out of concern for the quality of the food supply, mainly food adulteration and contamination issues that would first stem from intentional contamination to later with chemical food additives by the 1950s.

The development of colleges and universities worldwide, most notably in the United States, would expand food chemistry as well with research of the dietary substances, most notably the Single-grain experiment during 1907-11.

WATER

A major component of food is water, which can encompass anywhere from 50% in meat products to 95% in lettuce, cabbage, and tomato products. It is also an excellent place for bacterial growth and food spoilage if it is not properly processed.

One way this is measured in food is by water activity which is very important in the shelf life of many foods during processing. One of the keys to food preservation in most instances is reduce the amount of water or alter the water's characteristics to enhance shelf-life. Such methods include dehydration, freezing, and refrigeration.

CARBOHYDRATES

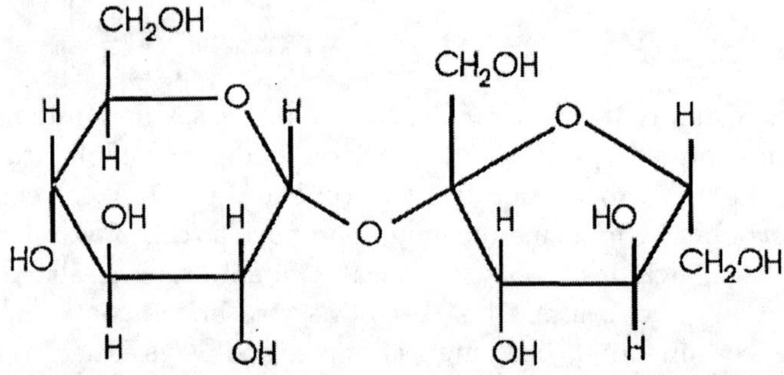

Sucrose: ordinary table sugar and probably the most familiar carbohydrate. Comprising 75% of the biological world and 80% of all food intake for human consumption, the most common known human carbohydrate is starch.

The simplest version of a carbohydrate is a monosaccharide which possesses the properties of carbon, hydrogen, and oxygen in a 1:2:1 ratio under a general formula of $C_nH_{2n}O_n$ where n is a minimum of 3. Glucose is an example of a monosaccharide as is fructose.

A chain of monosaccharides form to make a polysaccharide. Such polysaccharides include pectin, dextran, agar, and xanthan. Sugar content is commonly measured in degrees brix.

LIPIDS

The term lipid comprises a diverse range of molecules and to some extent is a catchall for relatively water-insoluble or nonpolar compounds of biological origin, including waxes, fatty acids (including essential fatty acids), fatty-acid derived pHospHolipids, spHingolipids, glycolipids and terpenoids, such as retinoids and steroids. Some lipids are linear alipHatic molecules, while others have ring structures. Some are aromatic, while others are not. Some are flexible, while others are rigid.

Most lipids have some polar character in addition to being largely nonpolar. Generally, the bulk of their structure is nonpolar or hydropHobic ("water-fearing"), meaning that it does not interact well with polar solvents like water. Another part of their structure is polar or hydropHilic ("water-loving") and will tend to associate with polar solvents like water. This makes them ampHipHilic molecules (having both hydropHobic and hydropHilic portions). In the case of cholesterol, the polar group is a mere -OH (hydroxyl or alcohol).

Lipids in food include the oils of such grains as corn, soybean, from animal fats, and are parts of many foods such as milk, cheese, and meat. They also act as vitamin carriers as well.

PROTEINS

Proteins compose over 50% of the dry weight of an average living cell and are very complex macromolecules. They also play a fundamental role in the structure and function of cells. Comprised mainly of carbon, hydrogen, oxygen, and some sulfur, they also may contain iron, copper, pHospHorus, or zinc.

In food, proteins are essential for growth and survival and vary depending upon a person's age and pHysiology (e.g., pregnancy). Proteins in food are commonly found in peanuts, meat, poultry, and seafood. They are also involved in ELISA test for food allergy determination as well.

ENZYMES

Enzymes are biochemical catalysts used in converting processes from one substance to another. They are also involved in reducing the amount of time and energy required to complete a chemical process. Many aspects of the food industry use catalysts, including baking, brewing, dairy, and fruit juices, to make cheese, beer, and bread.

VITAMINS

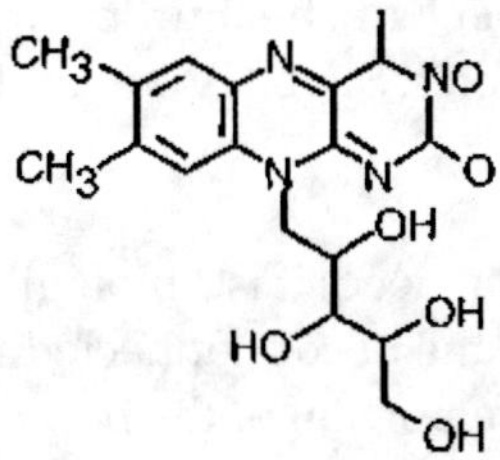

Riboflavin (Vitamin B2), water soluble

Vitamins are nutrients required in small amounts for essential-metabolic reactions in the body. These are broken down in nutrition as either water soluble (Vitamin C) or fat soluble (Vitamin E). An adequate supply of vitamins can prevent such diseases as beriberi, anemia, and scurvy while an overdose of vitamins can produce nausea and vomiting or even death.

MINERALS

Dietary minerals in foods are large and diverse with many required to function while other trace elements can be hazardous if consumed in excessive amounts.

Bulk minerals with a Reference Daily Intake (RDI, formerly Recommended Daily Allowance (RDA)) of more than 200 mg/day are calcium, magnesium, and potassium while important trace minerals (RDI less than 200 mg/day) are copper, iron, and zinc. These are found in many foods, but can also be taken in dietary supplements.

COLOUR

Food coloring is added to change the colour of any food substance. It is mainly for sensory analysis purposes. It can be used to simulate the natural colour of a product as perceived by the customer, such as red dye like FD&C Red No.40 (Allura Red AC) to ketchup or to add unnatural colors to a product like Kellogg's Froot Loops. Caramel is a natural food dye; the industrial form, caramel coloring, is the most widely-used food coloring and is found in foods from soft drinks to soy sauce, bread, and pickles.

FLAVORS

Flavor in food is important in how food smells and tastes to the consumer, especially in sensory analysis. Some of these products occur naturally like salt and sugar, but flavor chemists (called a "flavorist") develop many of these flavors for food products. Such artificial flavors include methyl salicylate which creates the Wintergreen odor and lactic acid which gives milk a tart taste.

FOOD ADDITIVES

Food additives are substances added to food for preserving flavors, or improving taste or appearance. These processes are as old as adding vinegar for pickling or as an emulsifier for emulsion mixtures like mayonnaise. These are generally listed by "E number" in the European Union or GRAS ("Generally recognized as safe") by the United States Food and Drug Administration.

SCIENCE AND FOOD

FOOD is not merely a social need: it is a biological one. All the same, as society grows in size and complexity, the needs connected with food get more varied and more elaborated. Today a large portion of the population live in big towns, so that the transport and distribution of everyday articles of diet, especially perishable ones like milk or fish, have become a serious problem.

Today, too, people want food from all over the world think of West Indian bananas, New Zealand mutton, Californian grape-fruit, and Argentine beef. And this involves even bigger problems of transport and distribution. Then, with the growth of transport facilities, the natural tendency has been for food to be grown intensively, often on an enormous scale as on the Canadian wheat-farms or the South American cattle- ranches in the regions of the world best suited to its production, and then exported. Food production is now largely a capitalist business enterprise carried out on a world scale. And this, of course, has sharpened competition. The result is that there is a premium on certain kinds of improvements at every stage of the business the making of new varieties of food-plants and food-animals, the technique of growing and rearing them, their storage and preservation in transit; and at every point science has been called in to help in making these improvements.

Mendel's work was published in 1865, and lay unnoticed between the covers of a rather obscure scientific journal for thirty-five years. When it was unearthed again in 1900, it was immediately seen to be of outstanding importance to the science of heredity. Among the men who saw its importance was William Bateson of Cambridge, and it was through his insight and enthusiasm that Cambridge soon became a centre for research in this new branch of science.

The essence of Mendelism is that hereditary characters are determined by definite unit particles which are handed on from parent to offspring in the reproductive cells, and that, owing to the microscopic machinery of those cells, the hereditary units can be shuffled and re-dealt in new ways. This is quite contrary to the older ideas of heredity, which usually assumed that the characters of the parents became permanently blended in the offspring, more or less as coloured ink blends with water. Mendelism showed that there was no blending, but that, by means of properly-planned crosses, the various characters of different races can be taken to bits, so to speak, and reassembled almost at will in all kinds of new and pure breeding combinations.

For obvious practical reasons of space and time, most of the early work on this new branch of science was done on small and, if possible, quick-breeding species—mice, primroses, flies, sweet peas, and the like—irrespective of their usefulness or uselessness to man. But soon it became clear that Mendel's laws were general, and that they applied to every variety of character in every kind of animal and plant; and then, naturally, scientists began to try to apply their knowledge to practical ends.

Thus Biffen's very practical work grew naturally out of the pioneer studies of Mendel and Bateson. If they were right, then it should be possible to build up new varieties of plants and animals by deliberate crosses, instead of confining yourself to slow selection just as modern chemistry deliberately makes new substances by basing itself on the atomic theory, instead of mixing things in hit-or-miss fashion in a test-tube. This is what Biffen set himself to do with wheat.

Let me take one example of his work. He aimed at combining resistance to rust disease, which he found in an otherwise poor strain of wheat, with the high yield of one of the best cropping strains. So he crossed the two: —and the offspring were all susceptible to disease! This would have been discouraging to the older breeders—but not to a Mendelian, who knows that characters can be masked for a generation, but can still be bred out pure in later generations. In his crosses he introduced also the character of "hard" grain, rich in gluten, which the millers asked for, and strong straw to prevent the plants from being too easily "laid" by storms, and succeeded in combining the characters he wanted after several generations of planned crossing and selection.

The two main strains he manufactured thus were called Little Joss and Yeoman. Although not put on the market till 1912 and 1917 respectively, by 1927 they occupied about a third of the entire world's wheat lands.

However, Yeoman does not suit all soils, especially clays and light sands; and now one of Biffen's chief aims is to make new types of Yeoman—by breeding in new features from other strains—which will suit every kind of wheat land. It is on this problem that he is mainly busy. Besides that, he is trying to manufacture spring wheat suitable for this country.

Spring wheats are not necessary with our mild winters; but the introduction of sugar-beet, which is not harvested until November or December, has made it desirable to have spring wheat to put on land that has been under beet the previous year. Bull-dog calves are a form of monstrosity which occurred with increasing frequency in the Dexter breed of cattle: they sometimes occurred to the tune of 20 or 25 per cent of all births, and some breeders were giving up their herds. Crew undertook to analyse the matter on Mendelian principles. He showed that you were bound to expect a high percentage of these monstrosities if you followed certain principles of breeding. What is more, he showed that if you followed certain other principles, you would no longer get any "bull-dogs" born.

The investigation on milk-production began with an examination of all the herd-books which the Department could get from breeders—a task which obviously only a big central institution could undertake. The milk records of thousands of cows were studied, and also their pedigrees.

This showed, as clearly as any paper analysis without actual breeding experiments can show, that the tendency to high milk-production was due largely to what scientists Call sex-linked inheritance, which is a special case of Mendelian inheritance. A sex-linked factor means something which a father transmits to all his daughters, but to none of his sons. A male, on the other hand, can only receive it from his mother. A well-known example from human beings is excessive bleeding or haemopHilia -the disease which the Russian Tsarevitch had. He got the hereditary factor for bleeding through his mother, and if he had lived and married, could only have passed it on through his daughters.

Most hereditary factors, of course, are transmitted equally from father or mother to all offspring of either sex. You will readily see that if a sex-linked factor enters into milkproduction, breeders must use quite a different system of breeding from the ordinary. For instance, if you find a bull which sires a large number of high-yielding cows, instead of keeping his sons to breed from, as you would naturally expect, you must reject them all—for none of them will have inherited the father's good qualities.

What Crew would like to do is to test this out by actual experiment; but this would take years and be pretty costly, and at the moment he cannot get the money for such an experiment. However, it is clear, I think, that he is on the track of something which could put up the average milk-yield of the cows of this country by anything from 20 to 40 or perhaps even 50 per cent.

It is not enough to breed new varieties which seem all right on the experimental plot. They must be tested out under commercial conditions before they can be safely recommended to farmers. Biffen's own department cannot very well do this, and as a matter of fact the business is entrusted to the National Institute of Agricultural Botany.

Testing is just as important in its way as breeding, for each variety will behave differently in different conditions of soils and climate; and the Institute, with its six testing stations in different parts of the country, helps with this. It would be a good thing if there were more stations, covering the country still more thoroughly, but meanwhile the Institute supplies this necessary link in a pretty adequate way.

Then we must not forget the soil. There are a few soils which are rich enough to dispense with chemical treatment, but with most soils you must put fertilizer on if the plant is to take out all it needs for optimum growth. The fundamentals of this branch of science are studied at the big Experimental Station at Rothamsted, directed by Sir John Russell, and many detailed soil problems are being worked out at the Experimental Farm belonging to Imperial Chemical Industries at Jealott's Hill.

I have not space to say much about this work, except that it is extremely important. It is only by careful studies, year after year, that we can find out how to bring land to its full yielding capacity and keep it there. And our present knowledge, if properly applied, would make it possible to bring off much bigger crops, especially with plants other than wheat, —such as oats or potatoes; and also with grass. But grass is animal food, not human food, and that brings me back to the question of how science can help with live-stock raising.

With regard to animals, the problem is more complicated than with plants, because there is an extra link in it—namely, the plants which the animals eat. Let us look at some of the points that arise. You may have the most wonderful breed of cattle or sheep in the world; but if you put them on poor pasture they will not do well. In fact, on really poor pasture they will do a good deal worse than much inferior stock, because to live and grow at all they need more and better food than the mediocre beasts.

The pasture, in its turn, may be poor because it consists of poor kinds of grasses, or because it is growing on soil which is deficient in some important substance, like iron, or lime, or pHospHorus. Further, the soil may be naturally deficient in the substance—for instance, there are big areas of the middle west of America where (owing to the ice sheets of the glacial period) there is practically no iodine in the soil, and therefore human beings are unusually prone to develop goitre, and pigs are often born hairless and soon die. Or man may have caused the deficiency by constantly taking stuff out of the soil in the shape of the lime and pHospHorus in the bones and meat of his animals, the iron in their blood, and so on, and never putting any back in the shape of fertilizer. Two splendid Institutions I visited are especially concerned with these problems—the Welsh Plant breeding Station at Aberystwyth and the Rowett Institute for the Study of Animal Nutrition at Aberdeen, both of them largely financed from State funds. Let me begin with the latter. From its start in 1922 the work here has been directed by Dr. John Orr, and has had the general purpose of finding out the relation between diet and disease, both in animals and in men.

The researches started on these lines have not only revealed how to cure a number of obvious diseases, but have also shown that many animals and people who could never be classified as diseased are really suffering from slight deficiencies of diet, and that their health and vigour can be increased—often startlingly increased if the diet is corrected. So that, instead of the rather negative idea of remedying obvious disease, the positive aim of promoting health through diet is gradually becoming dominant in the Institute.

I will only give one main sample of the work done here— the kind of laborious work which obviously can be undertaken only by a national institution with a long-term policy and regular funds at its disposal. It concerned the nature of the diet afforded by different pastures to the live-stock of this island. To start with, a pasture survey was made. Samples of herbage from nearly 400 different localities were taken and carefully analysed chemically. One thing that early emerged was that almost any pasture, even the best-looking, could be improved both in the quantity and quality of its yield by adding the proper fertilizers.

The next and perhaps more important point was that the great bulk of the hill pastures, which occupy such enormous tracts of our country, especially in Scotland and Wales, were badly below standard in regard to the amount of lime and pHospHorus in the herbage they grow; and further, that this lack of the minerals necessary for bone-growth and for health in the animals pastured on them was almost always linked up with a deficiency in the nitrogen needed to build up flesh.

It was further pretty clear that, in general, low mineral content of the pastures went hand in hand with a high incidence of disease in the stock-mostly sheep-which grazed them; but to get accurate information on this, the Institute found it necessary to buy a farm of its own in Argyllshire, where careful experiment has been going on for nearly four years. This farm, in the mineral content of its herbage, is near the average of hill pastures.

The experiments have shown that sheep on this farm not only did not get enough lime and pHospHorus for healthy growth at any time of the year, but in the winter months were not getting enough sheer nourishment, as measured in the energy-units that pHysiologists call calories. In winter-time, to get full health and reasonable growth, it was necessary to supplement the herbage with extra nutriment such as maize, and also extra minerals—or else to treat the soil beforehand with a fertilizer which would supply the deficiencies.

Further, it appears not only that a large area of the Scottish hill pasture is deficient in lime and other vital substances, but also that the deficiency has been getting steadily worse for the last fifty years or so. It seems clear that by a proper use of mineral fertilizers on the pastures, or by extra mineral containing rations for the stock, the carrying capacity of hill pastures could be at least doubled—that man could make two sheep grow where only one grew before.

As a startling example of what proper feeding and careful management can do, I might mention that this year a cow at the Institute was made to rear no fewer than eleven calves in a single milk-period, in place of the two to four which is the usual number!

Professor Stapledon is an enthusiast about pasture plants in general and grasses in particular. I should imagine that he knows more about grasses and clovers than any other man in the world. I have not the time to do more than just mention the special bits of research that are going on at his Institute: an experiment with red clover involving half a million artificial crosses between different strains; work which is aimed at providing a new type of white clover better than the ordinary Dutch strain; breeding and selection experiments with oats and different kinds of grasses.

These last, by the way, have an interest of a general kind. Most researches of the sort set out to breed a variety which will give bigger yield or better growth in the most favourable conditions. But Stapledon, with the Welsh hills all round him, has his eye on the uplands, and his work is to select for unfavourable conditions—to make it possible for a richer kind of grass to grow in place of the usual moorland bents.

But the most obviously exciting work is going on high up in the mountains. Long years of careful research on a comparatively small scale had convinced Stapledon that he could change the whole character of the hill vegetation for the better. Last year Sir Julien Cahn gave a considerable sum of money to put these ideas to the test on a large scale. With this money Stapledon purchased two tracts of land, one between 900 and 1300 feet up, another bigger one above the 1500-foot level. Both are just rough mountain pasture, yielding only a scanty nourishment to sheep even in summer. In winter the flocks from such areas have to be sent down and boarded out at so much a head on the pastures of lowland farms.

At the time of my visit at the end of September, the upland pastures were all turning ashen brown. Here and there, however, areas of summery green showed in the autumnal landscape. These were patches which Stapledon had treated according to his methods only a short six months before. They were covered with typical lowland grasses, which continue leafing much later than the types adapted to the bleak moorlands, and with a good proportion of clover.

Briefly, Stapledon has discovered that practically any hillside, at least up to the 2000-foot level, can be turned into pasture of lowland type by combining three procedures. The first is to get rid of the existing vegetation and to break up the soil; the second is to sow with the right mixture of grass and clover seeds, and the third is to supply the right blend of mineral fertilizers. Once this is done, proper grazing, with the occasional addition of fertilizer, will keep the pastures in condition.

Each of these three procedures has its long background of scientific and technical research. The first would be impossible without the development of the caterpillar tractor in the last ten years or so, to pull agricultural implements over otherwise inaccessible hillsides. The right kind of tractors is already available; but the right kind of implements is now the practical core of the problem.

Ploughing would do, but is expensive anyhow, and the ploughshares often get broken on stones. Harrowing is another method; scraping, as for aerodrome surfaces or road-grading, another. Up in one corner of the moor is a collection of different types of implements that are being tried out. Perhaps a special one will have to be designed. But the finding of the right method here is now merely a matter of time and experiment.

Then there is the second procedure—of sowing the right seed mixture. The background to this is long experience, and also years of rigorous selection, by which Stapledon has manufactured types of lowland grasses which will stand up to hill conditions.

And back of the third procedure—of supplying the right fertilizer—is nearly a century of experiment, beginning with the pioneer work of Lawes and Gilbert at Rothamsted in the early Victorian era. None of the procedures alone, or in couples, will do. All three are needed to establish the richer and more lasting pasture.

The lowland type of pasture thus established has all sorts of advantages. For one thing, it is more nutritious in itself; and, perhaps more important, it begins providing nutriment much earlier in the year and goes on providing it much longer. With such pastures available, the hill farmers would be able to graze many more sheep on the same land, and to keep them all the year round, instead of incurring the trouble and expense of boarding them out in the lowlands in winter. And there would be other advantages concerned with rearing fat lambs for sale at the best time for the market.

When you consider that nearly a quarter of the area of Great Britain consists of mountain and rough hill grazings, you will realise what a staggering change could be made by putting the results of these researches into practice. Even if the results were only applied up to 1500 feet, the change would be enormous. And now I ought to say something as to the general ideas which I got from what I have seen. In the first place, the fundamental research is, without doubt, well organized and ably directed. In plant and animal breeding, grass research, animal nutrition, soil science, fisheries, food storage, the study of the animal and plant enemies and parasites of crops, our British laboratories, such as those at Cambridge, Edinburgh, Aberystwyth, Aberdeen, Hull, Rothamsted, South Kensington, Jealott's Hill, backed up by the agricultural colleges and the University departments concerned with agriculture, are doing work of outstanding quality—original and vitally important. But practically all the research men I have talked to showed an interesting mixture of optimism and pessimism. They all knew the scientific importance of their own work, and were convinced of its possible value for practice; several of them said to me that a doubling of the present amount of food grown in this country was not only perfectly possible, but a modest estimate of what could be achieved by applying the scientific knowledge which exists. That, by the way, would make these islands self-supporting in regard to most foodstuffs (though not wheat).

But in contrast with this optimism as to possibilities was certain pessimism as to actualities. What is the good of doubling the number of sheep in the country if sheep prices may fall so low as to wipe out any reasonable profit to the farmer? What is the good of inventing new brands of wheat that will make it possible to grow more bushels of wheat to the acre or to push wheat cultivation nearer the pole, if the world's wheat producers have on their hands vast surpluses they cannot dispose of profitably and are clamouring for a restriction of output and cultivation? What is the good of inventing new methods of cold storage which enable ships to ransack the recesses of the Arctic Ocean for fish, if a large proportion of the annual catch is thrown away or disposed of for manure?

Then of course, there is the conflict between home and empire. What about New Zealand mutton and Australian beef if we double our own livestock, or Canadian wheat and apples if we increase the home output? And there still remains the problem of the balance between agriculture and industry. If we are to go on exporting coal and steel and machinery and motor-cars to other countries, they must be paid for by our taking something in exchange from those countries: foodstuffs are one of the biggest items in that balance-sheet.

Then there are the conflicts of interests at home. There is the interest of the breeder of pedigree cattle of a certain breed against the general livestock interest; that of the seedsman as against the farmer; of the wool merchant against the sheepbreeder; of the middle-man as against the producer and the consumer-one could go on almost indefinitely!

The outlook, mind you, is not all black. Much of the scientific results could be applied to-morrow to bring down cost for. the farmers, and so make it possible for many more of them to make a decent assured profit, instead of scraping along in a hand-to-mouth way, or even falling over the edge into failure. But that is looking at the problem from the producer's angle only: the consumer is another matter.

That, however, is not the fault of science, but of our economic system, and how that is to be remedied is a question for economists and administrators. Meanwhile it does lie like a barrier across our hopes for a well nourished, healthy nation. But the hopes are there, and if the barrier of the economic system seems strong, the basis of the hope is strong too! And that basis is the certitude that science, if its existing knowledge were properly applied, could at least double the amount of food we produce in these little islands, and could put up world-production to a level at which there would be enough and to spare for the 2000 million human beings in existence.

CHAPTER 2

Biochemistry of Vitamins

The making of the vitamin concept in the first few decades of this century relied heavily on functional ascriptions, centred on the role of specific, but chemically ill-defined nutrients in preventing specific dietary deficiency diseases. In the interwar period, efforts were made increasingly to characterize vitamins in terms of their chemical structure. In the course of these structural pursuits, vitamins were turned from hypothetical explanatory devices into molecules that could be isolated, identified, manipulated and synthesised by scientists. In the process, standard quantitative requirements for individual vitamins were defined, which provided a new rationale for nutritional and medical intervention aimed at preventing and treating diseases.

Methods to produce vitamins in increasingly pure form were developed and exploited by industry, not only to manufacture standardized vitamin preparations for clinical use, but to sell vitamin preparations and vitamin-fortified foods to consumers. The promotion of vitamin supplements, in standardized and chemically defined form, served to persuade millions of people to buy these products for the benefit of their health.

The elucidation of vitamin structure came to be seen as an important scientific problem by different constituencies, with a focus on Britain. I shall draw attention to a number of scientific, medical, social, and industrial motivations which stimulated structural studies of vitamins, and outline how these different motivations were interlinked and mutually reinforced. My empHasis here being on the multiplicity of reasons underlying structural pursuits, I shall be less concerned with the detailed scientific steps whereby vitamins were turned into well-defined molecules. Several issues of policy and strategy that will be discussed encompassed a much broader domain than vitamin research. On that level, the vitamin case is presented to illustrate some general features of the dynamics of molecularization of biology and medicine in the interwar period.

I shall begin by discussing how vitamins acquired political prominence in Britain during World War I. The social and scientific importance ascribed to vitamins at this time informed government food policy, boosted vitamin research and turned vitamins into a matter of concern for the food industry. As I aim to show, the wartime formation of links between these domains provided a major impetus for the molecularization of vitamins after the war. I shall discuss the support of the Medical Research Council for vitamin research in interwar Britain, and tensions which arose between the Secretary of the Medical Research Council, Walter Morley Fletcher, and the biochemist whom Fletcher expected to take a leading role in investigating the metabolic functions of vitamins, Frederick Gowland Hopkins.

I shall argue that Hopkins came to use vitamins for his own ends, to promote biochemistry as a fundamental science of life, at the same time reinforcing a molecular perspective, not only of diet, but of all living processes. I shall then discuss how vitamins motivated the British food industry to transform its practices of quality control and product promotion and development, by recruiting research scientists. Finally, I shall point to the ways in which the different interests which motivated vitamin research in the interwar period were interconnected by, and converged on, a concern with vitamin structure.

THE POLITICS OF VITAMINS

In World War I, concern with the food supply and the threat of food shortages focussed political attention on the composition of a healthy diet, in relation to both army rations and foods available to the civilian population. In Britain especially, scientists were involved in food policy formulation at the highest levels, and they drew on new understanding of the importance of the qualitative composition of the diet, over and above its quantitative energy content measured in calories.

At the same time, nutrition research to further this understanding was given high priority. Moreover, the political and scientific networks established in conjunction with these efforts, especially through the involvement of the Medical Research Committee, were of enormous influence in guiding scientific research on nutrition after the war. "Accessory food factors", or vitamins as they became known later, were at the centre of these developments.

In the decade before the war, animal feeding experiments with synthetic diets of known chemical composition had indicated a vital nutritional role for hitherto unsuspected components of the diet. The demonstration that animals cannot survive on a diet of pure carbohydrates, fats, proteins, water and minerals, but that small supplements of, for instance, milk restored normal growth, was taken to indicate the presence in normal dietaries of as yet unidentified "accessory food substances" that are essential components of the diet.

Scientific interest in these hypothetical dietary substances was heightened when they were linked tentatively to diseases that were beginning to be ascribed to dietary deficiencies. The demonstrated role of specific foods in preventing and curing diseases such as beri-beri and scurvy stimulated efforts to isolate and characterize the protective components in these foods. At the Lister Institute in

London, for example, attempts to identify the protective ("antineuritic") factor implicated in beri-beri heralded the start, in 1910, of a renowned long-term programme in nutrition research. As part of these attempts, an animal assay was developed to determine the quantitative distribution of the protective factor in different foods. This experience was built upon during the war, when the Royal Army Medical Corps sought the advice of the Lister Institute in the face of outbreaks of beri-beri and scurvy among British troops in the Middle East and India.

An animal assay for antiscorbutic activity was developed in addition to that for antineuritic activity, and extensive experiments were done to determine which foods, in a form suitable for transport to overseas troops, are most effective in preventing scurvy and beri-beri. Many of the male scientists being away at the front, most of this research was done by women scientists (many of them newly appointed for this purpose), led by Harriette Chick. Through this work, Chick and her women colleagues acquired a high scientific profile in nutritional research, which was further enhanced after the war. Nutritional work such as that at the Lister Institute had the active support of the Medical Research Committee as well as the Food Committee appointed under the Royal Society Committee upon the subdivision, in 1916, of its Sectional Committee on Physiology. The Royal Society Committees, with memberships drawn from the most reputable scientific circles, were set up to organize war-related research and to advise government on policy matters.

The Chairman of the Food Committee, eminent pHysiologist E.H. Starling, was, in that capacity, also Honorary Scientific Advisor to the newly created Ministry of Food. The Committee's Secretary was Walter Morley Fletcher, former Cambridge pHysiologist and Secretary of the Medical Research Committee since its establishment in 1913. Among the Food Committee's members was Cambridge biochemist Frederick Gowland Hopkins, by then highly esteemed for his innovative work on the pHysiological role of essential amino acids in dietary protein and of chemically undefined accessory food factors in the diet.

Hopkins' presence on the Committee, and the high regard in which his work was held by fellow members, ensured that accessory food factors received full attention. At the level of food policy, the Committee placed the empHasis on the importance of fresh foods, in the light of evidence suggesting that processed foods such as mechanically milled rice, white bread, dried milk and canned meats and vegetables are deficient in dietary factors essential for health. This empHasis on fresh foods, emanating from such an authoritative body, created concern among food manufacturers about adverse effects on sales of their allegedly inferior products and motivated them to set up their own research laboratories.

At the level of research, the Food Committee gave high priority to the determination of the accessory food factor content of different foods as measured by biological assays, but issues of etiology also received attention. In particular, the Medical Research Committee funded investigations of the etiology of rickets from different angles, with Edward Mellanby focussing on the possible dietary connection in feeding experiments with puppies.

His results provided evidence for a causal link with a fat-soluble accessory food factor in the diet, but this interpretation was opposed fiercely by a group of pHysiologists in Glasgow, also funded by the Medical Research Committee, on the basis of their field studies among the urban poor. In fact, the rickets controversy was to continue acrimoniously throughout the 1920s, and the disease became a focus for scientific research and experiments in public health.

The Medical Research Committee's active involvement in the scientific war effort had important repercussions after the war, when it was transformed into the Medical Research Council (MRC). The Committee had been set up initially for the promotion, coordination and support of tuberculosis research, according to provisions made under the 1911 National Insurance Act. Its intertwinement with the Royal Society Committee, however, gave it a much broader role in defining medical research policy.

Initially an ad hoc arrangement contingent on wartime conditions, this role was formalized after the war with the creation of the MRC. For the first time in Britain, there was a government funding body for the whole range of medical research, endowed with a large budget and responsible directly to the Privy Council rather than the Ministry of Health. As Secretary of the MRC, Fletcher gained enormous powers in the direction of medical research in Britain, and he built on the structures put into place during the war. Those serving on the MRC included powerful allies of Fletcher's among the scientific establishment, who shared and informed his objectives. Among Fletcher's objectives was his ambition that the MRC should become the driving force behind fundamental medical research, with a view to building strong scientific foundations for clinical practice. This aim was in harmony with Fletcher's own scientific background.

In the Physiological Laboratory in Cambridge, he had worked on fundamental aspects of muscle pHysiology, in an institution which, under the leadership of Michael Foster, had fought hard for the autonomy of its research programs, independent of immediate medical concerns. Furthermore, many of Fletcher's closest advisors on the MRC were themselves engaged in basic research and intent on replacing supposedly subjective clinical judgement with supposedly objective laboratory research as the court of authority for medical matters. While Fletcher promoted fundamental research in pHysiology, biochemistry and pathology without demanding short-term medical applications, he nevertheless empHasised that clinical applications should be the long-term goal of this research.

One of the areas privileged by the MRC was vitamin research. Not only was this subject deemed to be of the greatest medical and social importance, wartime efforts to coordinate research in this area had already resulted in the establishment of a special advisory committee for the promotion of research into the nature and role of accessory food factors. In 1918, the Medical Research Committee and the Lister Institute had jointly established the Accessory Food Factors Committee, with Hopkins in the Chair, Harriette Chick as Secretary, and Jack Drummond, Arthur Harden and Edward Mellanby as members.

This committee remained an integral part of the MRC into the 1930s and played a central role in defining the directions of MRC-supported vitamin research. It also issued a set of highly authoritative and widely read reports on the state of knowledge of vitamins.

An early post-war opportunity to demonstrate the medical and social importance of vitamin work, and of the MRCs beneficial role in supporting it, presented itself with the widespread famine in Central and Eastern Europe which followed the war. Diseases such as scurvy and rickets were rife, especially among children. Under the auspices of the Accessory Food Factors Committee, Harriette Chick and Margaret Hume, later joined by more women Colleagues, went to famine-struck Vienna in 1919, to investigate whether the diseases affecting the population were due to vitamin deficiencies. For three years, they studied the effects of specific dietary supplements on the prevention and cure of rickets, at a time when the etiology of this disease was still deeply contentious. The work was funded jointly by the MRC and the Lister Institute, and Chick's team was offered research facilities in the University of Vienna's children's clinic by its Professor of Pediatrics, Clemens von Pirquet.

In his clinic, Pirquet had introduced a rigorous dietary system, whereby the daily diet of individual children was related quantitatively, in terms of protein and calorie content, to each child's size. Under these conditions, it was possible to conduct controlled trials on human subjects, equivalent to systematic animal feeding experiments. By giving different groups of children different or no supplements to their controlled diet, Chick and her colleagues established convincingly that rickets can be prevented and cured by cod-liver oil. The results were sufficiently dramatic to lend credence to the vitamin deficiency theory for many of those who had previously regarded rickets as an infectious disease, including Pirquet himself.

In this instance, therefore, the active day-to-day involvement of research scientists in clinical practice served to persuade initially sceptical clinicians of the significance of vitamins, by means of clincial demonstration. Although the etiology of rickets nevertheless remained a contentious issue, this clinical success immediately became a favourite example used by those promoting the importance of vitamins and vitamin research.

Dietary intervention in the light of the vitamin theory thus yielded impressive therapeutic successes from an early stage onwards, notably in the case of the putative vitamin deficiency diseases-beri-beri, scurvy and rickets-associated strongly with poverty, famine and war. Such success in the treatment and prevention of socially and politically significant diseases not only raised the profile of vitamins in public health policy and strengthened the vitamin theory, but also stimulated a large scientific research effort related to vitamins.

THE SCIENCE OF VITAMINS

In interwar Britain, vitamin research was supported and promoted above all by the MRC and its Accessory Food Factors Committee. The MRC funded investigations of the etiology of deficiency diseases; the development and perfection of quantitative biological and chemical assays of vitamin activity, with a view to determining minimum requirements and setting international standards; studies of the chemical properties and structure of vitamins; and the pHysiological role of vitamins. I here want to sketch how these different fields of investigation had bearing on each other, with structural questions becoming the linking theme.

Questions about the chemical nature and the pHysiological role of vitamins had been interlinked from the very beginning, with the formulation of the notion of accessory food factors to explain the results of animal feeding experiments. Using controlled diets of systematically varied chemical composition, Hopkins showed that rats stop growing and die if they are fed an otherwise adequate diet of pure protein, carbohydrate, fat, water and minerals, while normal growth is restored by the addition of small quantities of milk.,

On these grounds, he postulated the existence in normal diets of "accessory food factors". These hypothetical substances appeared to be necessary for normal growth and survival, but they had never been detected in chemical analyses of foods, let alone identified chemically. Presumably, therefore, accessory food factors were present in minute amounts in normal dietaries. In order to consolidate the notion of accessory food factors, and to investigate their precise pHysiological role in metabolism, it became imperative to isolate and identify these substances. Hopkins himself tried to do so for several years, without success.

Hopkins was more successful in the identification of "essential amino acids" present in dietary protein, that is, amino acids which are required for normal metabolism, but cannot apparently be synthesised in the body from other substances and, hence, must be supplied in the diet. In fact, Hopkins had first become interested in the biochemistry of nutrition as a result of his demonstration with Edith Willcock that rats cannot survive on diets with adequate calorie content if the only protein source lacks the amino acid tryptopHan. He went on to identify and investigate several other essential amino acids. This work, together with the notion of accessory food factors, was instrumental in shifting attention from the quantitative calorie content of diet to its composition at the level of fine chemical detail. I shall come back later to the way in which this shift was used by Hopkins in the promotion of a molecular view, not only of diet, but of the living organism in general.

That food fractions providing protection against dietary deficiency diseases contain only the tiniest amounts of protective factors is illustrated by the studies of the antineuritic factor implicated in beri-beri. After a series of laborious extractions, evaporations and other chemical operations, Funk obtained just 0.5g of active substance from 54 kg of rice polishings. He identified his crystalline preparation as an organic base of the amine class and introduced the term "vitamine" to describe substances required to prevent "vitamine deficiency diseases", under which he categorized beri-beri, scurvy, pellagra, epidemic dropsy and, tentatively, rickets. However, after further elaborate chemical experiments, Funk soon found that the "vitamine" which he thought he had purified was, in fact, an impure mixture. Among the accessory food factors community, including Funk's colleagues at the Lister Institute, this set-back served as a cautionary message about the dangers of making premature claims about the chemical structure of these elusive substances. The chemist's difficulties in isolating vitamins were compounded by their instability to heat, oxidation, drying, hydrolysis and chemical reagents.

Yet in the course of extensive chemical investigation in different centres, combined with animal feeding experiments, accessory food factors began to be differentiated on the basis of chemical properties. By 1920, three factors had been distinguished: those named "fat-soluble A" and "water-soluble B" by McCollum and Davis working at the University of Wisconsin, and, following similar terminology, "water-soluble C" by Drummond in London. In line with Drummond's suggestion, these became known as vitamins A, B and C, and were soon followed by vitamins D (the antirachitic factor) and E (then also known as the fertility factor). A whole string of vitamins was added in the 1920s and 1930s, in alpHabetical sequence up to vitamin P, while vitamin B was subdivided into several different vitamins, first into vitamins Bl and B2, but eventually into twelve factors.

One important incentive for studying the chemistry of vitamins was linked to the determination of the differential distribution of individual vitamins in food-stuffs, in order to work out healthy diets defined in chemical terms. Increasingly, animal experiments showed that a mildly deficient intake of vitamins did not necessarily or immediately lead to severe clinical symptoms, such as those associated with scurvy, rickets or beri-beri, but nevertheless produced abnormalities in certain tissues, organs or structures of the body which affect health in the long term. In order to define accurate standards for minirrfum vitamin requirements to prevent disease and optimal vitamin requirements to ensure good health, pure vitamin preparations were required.

Standardization of vitamin requirements was one of the objectives of the MRC, but took on an international dimension through the activities of the League of Nations Health Organisation (LNHO). As Weindling has shown, the LNHO played an instrumental role in promoting optimal, instead of minimum, vitamin requirements. It was on the basis of international standards thus defined that John Boyd Orr concluded from his extensive investigations of dietaries that more than half the British population lived on an inadequate diet. By then, internationally agreed units of vitamin activity were measured, in bioassays, against standard samples, all held at the MRC's flagship institute, the National Institute for Medical Research in London.

By 1934, there were four of these standard preparations, by then in chemically well-defined form: vitamin A in the form of? -carotene; a standard absorption product of vitamin Bl; vitamin C as 1-ascorbic acid; and vitamin D in the form of calciferol, obtained by irradiation of ergosterol.

Chemical studies were also undertaken for etiological reasons. In the case of rickets in particular, there was the puzzling pHenomenon that not only cod-liver oil, but exposure to sunlight had a beneficial effect on patients. The effect of sunlight had been confirmed again in the investigations of Chick's team in Vienna, and while these investigations were in progress, Huldschinsky in Berlin cured rachitic children by irradiating them with a mercury lamp. At the University of Wisconsin, Harry Steenbock soon there after showed that many foods acquired antirachitic activity upon ultraviolet irradiation. This "Steenbock process" was patented by the University of Wisconsin. It was then shown, in Britain and the United States, that the irradiation of cholesterol produced small quantities of a substance with high antirachitic activity, presumed to be vitamin D.

This chemical conversion was further investigated by Otto Rosenheim's team at the National Institute for Medical Research, and by Adolf Windaus in Gottingen. These two groups, which exchanged information and materials, found that it was not pure cholesterol, but another sterol, separated from cholesterol with considerable difficulty, which produced the antirachitic activity upon irradiation. They named it ergosterol. Subsequently the substance produced by irradiation of ergosterol was isolated and named calciferol. Its remarkably high antirachitic activity was taken to indicate that this was vitamin D. The antirachitic effects of sunlight were attributed to the conversion of ergosterol ("pro-vitamin D") present in the skin into calciferol (vitamin D).

Windaus received the 1928 Nobel Prize in Chemistry for his work on the structure of sterols, including ergosterol. In fact, his structure of the sterol skeleton was later corrected, with Rosenheim's alternative structure receiving support from X-ray crystallograpHic studies by Bernal. Interestingly, it was only after the Chemistry Prize to Windaus that the Nobel Committee for Physiology or Medicine awarded their first Nobel Prize for vitamin work: in 1929, Hopkins and Eijkman shared the Prize for "the discovery of vitamins", as their work of several decades earlier was then interpreted.

Knowledge of the chemical structure of vitamins was also sought in order to gain insight into their mode of action in the body, and Walter Fletcher was particularly keen that the MRC should support fundamental research into the role of vitamins in metabolism. At first, vitamin function was viewed in global pHysiological terms: vitamins are required for normal growth, and vitamins are required to protect against particular nutritional deficiency syndromes. But functional questions typically became tied much more closely to questions about mechanisms: what was the mode of action of individual vitamins within the normal pHysiological process, what was the role of specific vitamins in metabolism?

Fletcher held that insight into these questions of function in a more precise, local sense would be of the greatest medical importance, both for the prevention and treatment of vitamin deficiencies by highly specific chemical intervention. If MRC- supported research were to yield such crucial results, it would simultaneously provide strong legitimation for the MRC's policy of supporting fundamental research, in this case in biochemistry. In the 1920s, Fletcher continuously needed to justify this policy to the clinically orientated medical establishment, however much he and his scientific colleagues favoured it and however much he asserted that it would ultimately yield important medical benefits.

Having the highest regard for Hopkins' earlier work on the biochemistry of nutrition, Fletcher expected that his former colleague in Cambridge and close ally on the MRC would adopt a leading role in biochemical research on the metabolic role of vitamins. To Fletcher's long-term frustration, Hopkins declined to fit in with this plan. In fact, Hopkins, disengaged himself from vitamin research altogether in the early 1920s. Ironically, in pursuing different, new priorities, he became a prominent propagandist for a molecular perspective on vitamins and diet.

Hopkins himself stressed repeatedly, in his writings and public lectures from the 1910s to the early 1930s, that investigations into the metabolic functions of vitamins could not even begin without knowledge of their chemical structure. This recurring message was consistent with Hopkins' own practices in biochemistry, where he always aimed to isolate and identify the molecules taking part in metabolic reactions, to elucidate the reaction mechanisms, and to investigate the control of these reactions by enzymes and their coordination within the cell.

Hopkins did not, however, encourage structural research on vitamins in his own department, with a view to investigating their metabolic role. Increasingly, research in the Cambridge Department of Biochemistry, which had been created in 1914, became focussed on fundamental biochemical research on biological oxidation and reduction and the catalytic control of specific oxidation-reduction reactions.

This shift in research was coupled strongly to Hopkins' ambition to transform biochemistry into a fundamental science of life that was to provide a unifying chemical perspective of all biological processes, across the living world. In particular, Hopkins' isolation in 1921 of glutathione, an agent with dual oxidizing and reducing properties which was found to be present in many different living organisms, stimulated the search in Hopkins' department for basic biochemical patterns common to a range of organisms, from bacteria to man.

"Fundamental" research had quite different connotations for Hopkins and Fletcher. While Fletcher promoted fundamental research with the aim of making medicine scientific, it was Hopkins' ambition to unite chemistry and all of biology. Thus, in the research pursued by Hopkins and his colleagues, medical questions, and indeed vitamins, became subordinate to a much more ambitious programme of molecularizing life, through biochemistry.

The Cambridge programme of biochemistry, however, remained atypical in the interwar years. Despite Hopkins' persistent stress on the underlying unity of this wide-ranging programme, this unity was not appreciated widely by influential British medical scientists, who generally regarded the Cambridge research as too far removed from human and animal pHysiology.

Hopkins countered these reservations by promoting his particular vision of life repeatedly in public. In order to give substance to that vision, however, he used therapeutic successes with vitamins and hormones as recurring illustrative examples of the social and medical need for the biochemist's special skills. In other words, Hopkins appealed to vitamins in his lectures and general writings for purposes of his own, to promote an approach to biochemistry which he defined as being concerned with "molecular dynamics in cells and tissues".

At the same time, Hopkins' frequently repeated message that knowledge of chemical structure was a prerequisite for the elucidation of the basic biological functions of vitamins (among other biological molecules) gave further prominence to the problem of vitamin structure. Hopkins' pronouncements mattered, because he was regarded as the pre-eminent public authority on the biochemistry of nutrition in interwar Britain, and played an active role in the most influential scientific bodies which shaped British science policy, including the MRC and the Royal Society.

While Hopkins himself did not practise what he preached with respect to vitamins, his department was the site where Albert Szent-Gyorgyi first isolated vitamin C. This work began in Cambridge as an unexpected byproduct of work on biological oxidation and reduction with Szent-Gyorgyi's efforts to isolate and characterize a substance with strong reducing properties from the adrenal cortex. In view of the pHysiological properties of this substance, identified as "hexuronic acid", there were strong indications that it might be identical with the antiscorbutic vitamin C. Assays with hexuronic acid, undertaken at Hopkins' insistence by S.S. Zilva at the Lister Institute, failed to support this hypothesis. Szent-Gyorgyi persisted in his investigations, however, and eventually, after his return to Hungary, produced much larger amounts of hexuronic acid from paprika and was able to confirm its identity with vitamin C. The substance was subsequently renamed ascorbic acid to give expression to its activity against scurvy.

It was the Cambridge research programme on mechanisms of biological oxidation and reduction that had attracted Szent-Gyorgyi to Hopkins' department. He later lamented the fact that his work on vitamin C had distracted him for years from his "fundamental" research in biochemistry, and described vitamins as being of greater interest to the chef than to the scientist.

A few months after these and other disparaging remarks were published, he was awarded, the 1937 Nobel Prize for Physiology or Medicine for the isolation, purification and biochemical characterization of vitamin C. In the same year, the Birmingham chemist Walter Haworth shared the Nobel Prize in Chemistry for his work on the structure and synthesis of vitamin C. His fellow laureate was Paul Karrer from the University of Zurich, who was honoured for his researches into the constitution of carotinoids, flavins and vitamins A and B.

In 1938, Richard Kuhn from Heidelberg University was awarded the Nobel Prize in Chemistry for his work on carotinoids and vitamins of the B group. (Kuhn was forbidden by the Nazi regime to travel to Stockholm to receive the award. These awards give an indication both of the difficulties involved in structural studies of vitamins and of the importance attached to the elucidation of vitamin structure by the scientific community.

The investigation of vitamin structure, then, was pursued for different ends: to determine detailed dietary requirements, to gain further insight into the etiology of vitamin deficiency syndromes, to initiate studies of their role in intermediary metabolism, and to produce work that could gain the highest scientific accolades in organic chemistry and biochemistry. There were also strong industrial incentives, to which I turn now.

VITAMINS AND INDUSTRY

Vitamins caused major concern to food manufacturers already during World War I, when assays of foods such as those performed at the Lister Institute suggested that many processed foods were deficient in accessory food factors. The empHasis which the Royal Society Food Committee placed on fresh foods in the food policies it advocated posed a serious threat to sales of processed foods and hence to profits of the food industry. To counter this threat, food manufacturers began to set up their own research laboratories after the war.

Here foods were assayed and analysed, and methods explored to enrich food products with vitamins obtained from natural sources. To keep production costs of vitamin-fortified foods low, and to develop the most convenient methods of production, the technical skills of chemists and nutrition scientists were put to use in the extraction, concentration, purification and chemical investigation of vitamins. The aim was to manufacture products that could be promoted and sold as having superior nutritional value to natural foods, especially with respect to their vitamin content. One of the outcomes was a role for the food industry in the molecularization of vitamins.

Manufacturers of margarines were among the first to be alarmed at allegations that their products, popular because of their low cost compared with butter, were lacking in the fatsoluble substances thought to promote growth (vitamin A) and to prevent rickets (vitamin D). The major British producer was the Maypole Dairy Company, but when Lever Brothers set up the Planters Margarine Company in 1914, the two manufacturers joined forces in research into the nutritive value of their products. In 1917 (before his move away from vitamin research), Hopkins began to act as a consultant for the two margarine companies and agreed to do research on their behalf, providing that he could publish his results. While very little is known about Hopkins' activities as a consultant to these companies, this early link between an academic authority on vitamins and industrial food research is of interest in itself.

Hopkins also had links in this period with two food companies in Norwich, Virol Ltd and Colman's, who advertised their products citing scientific findings about their vitamin content obtained in the Cambridge Biochemical Department. In 1924 Leslie Harris, who had completed his Ph.D. under Hopkins, went to work for Colman's, one of the first food companies to employ its own research scientists, and set up animal feeding experiments for purposes of quality control of proprietary baby foods. Harris was subsequently appointed head of the Dunn Nutritional Laboratory in Cambridge, founded with funds from the MRC and the Dunn Estate in 1927, when Fletcher had finally given up hope that Hopkins would return to vitamin research. This is one illustrative example of the new movement of nutrition scientists between industrial and pure research laboratories.

In the course of the 1920s, in the face of increasing competition among themselves and with the butter industry, British companies began to produce margarines fortified with vitamins A and D. Planters was the first to do so, in 1927, using vitamins extracted from liver oils. In 1929, The Industrial Chemist reported on independent tests done by the Pharmaceutical Society of Great Britain comparing the vitamin A and D content of butter and different margarines, using biological assays. The results for butter showed significant variation while the margarines displayed constant vitamin content and were superior in antirachitic activity to all samples of butter. The 1920s also saw the first production of vitamin preparations in Britain, both for clinical use and for public consumption as supplements to meals. The first ones were preparations of vitamin D produced by extraction from cod-liver oil, but this mode of production was superseded by the more convenient and cheaper method of producing vitamin D from irradiated ergosterol obtained from yeast. The first British product obtained this way was Radiostol, introduced to the market by British Drug Houses in 1927. It had an antirachitic activity 200,000 times higher, weight by weight, than that of codliver oil. The Director of British Drug Houses advocated both that this product be used by food manufacturers to fortify margarines, and that further chemical research be done to get insight into the chemical changes resulting from irradiating ergosterol.

In general, it was the pHarmaceutical industry that produced vitamin preparations, although British pHarmaceutical companies lagged behind German and Swiss companies in vitamin research and development. Not only was there a stronger tradition in organic chemistry in these countries, but the skills of organic chemists, both in structural analysis of natural products and in synthetic chemistry, had found a firm industrial base there much earlier than in Britain. In fact, the commercial production of synthetic vitamins in Britain began when the Swiss company Hoffmann-La Roche set up a "model plant" in Welwyn Garden City in 1937, for the industrial synthesis of pure vitamin C and Bl.

There was one British food company that did get involved in the production of vitamin preparations as well as vitamin-enriched foods and, through vitamins, was transformed into the pHarmaceutical company now known as Glaxo. Scientists working for the company were the motive force behind this transformation, which was coupled strongly to the molecularization of vitamins.

Glaxo grew out of JosepH Nathan & Company which produced dried milk and proprietary baby foods. In the face of poor sales of its milk powder sold under the brand name of "Defiance", the company relaunched this product in 1906 under the new name of "Glaxo". It was advertised as "the Food that Builds Bonnie Babies". The Nathan company's sales performance remained modest for some years, but a great boost to its fortunes came during World War I, as a result of a policy introduced by the new Ministry of Food, created in 1916. In 1917, the Ministry decided to provide the infant welfare centres of local authorities with milk powder, and the product it chose to distribute was Glaxo. After the war, greater competition made the company increasingly concerned about quality control of its dried milk products, and in 1919 it recruited Harry JepHcott, a young scientist with chemical and pHarmaceutical training. JepHcott was given the responsibility of ensuring for Glaxo a uniform and high standard of quality, and, because of the potentially serious commercial implications, he set out to investigate the then contentious issue of whether the drying process changed the vitamin content of milk. To help him in his investigations, JepHcott recruited Alfred Bacharach to the tiny scientific group at the Glaxo Department in 1920. Trained at the University of Cambridge, Bacharach had worked at the Wellcome Chemical Research laboratories from 1915 to 1918 and then at the Wellcome Chemical Works.

Together, JepHcott and Bacharach began a series of investigations into the antiscorbutic activity, in guinea pigs, of fresh milk and of milk dried by different processes. The results for the process used in the production of Glaxo compared favourably with other drying methods, but were not quite as good as those for fresh milk. This work was deemed sufficiently significant to be published in the Biochemical Journal. Indeed, from the early 1920s a good number of scientific papers was published in reputable scientific journals by people working in the Glaxo Research Laboratory.

In the light of the intensive research effort of the early 1920s on the dietary connections of rickets, JepHcott began to explore the possibility of producing a Glaxo powder fortified with vitamin D. In 1923 he attended the International Dairy Congress in Washington and took the opportunity to visit major nutritional centres at the University of Wisconsin and the University of Columbia, New York. At Columbia he met Theodore Zucker, who had perfected a process for extracting vitamin D from fish-liver oil. JepHcott secured the British licence for the Zucker process and in 1924 the Nathan company began to market vitamin D concentrate in pHials for drop dosage, under the name of Ostelin. This preparation, which had 2000 times the antirachitic activity of cod-liver oil, was the first standardized vitamin concentrate marketed in Britain. It was also the company's first pHarmaceutical product and it sold well, helped by recommendations in the medical press. Subsequently JepHcott obtained the licence for the Steenbock process for converting ergosterol into vitamin D by irradiation and, from 1929, Ostelin was produced in this way in the Glaxo Department using yeast as the source for ergosterol. The process had many advantages: the bulk buying of fish could be abandoned, yeast could be cultured relatively cheaply, uniformity of antirachitic activity could be secured more easily, and the product lacked the unpleasant fish taste of the earlier version of Ostelin.

A further change came in 1932, when the company began to produce pure crystalline calciferol from irradiated ergosterol. This product, Calciferol GL, was four times more active than the Steenbock irradiation product. In the 1930s, the Glaxo Department also developed a range of products containing vitamin A, it sold its vitamin products on an international scale in nearly all continents, and it produced a new proprietary infant food enriched with vitamins, called Farex. The commercial success of the company's vitamin products boosted the research department. In 1935, Glaxo Laboratories Ltd was incorporated as a private limited company, which was to "market proprietary pHarmaceutical products having reasonable claims to novelty". Gradually pHarmaceutical products became more important than food products as a percentage of net income. By the late 1930s, the company was doing research into the synthesis of vitamin A and iodine compounds, the isolation of bacterial antigens and pituitary hormones, and the isolation and extraction from liver of the agent that is active against pernicious anemia (later called vitamin B12). In conjunction with these research efforts, the scientific reputation of the Glaxo Laboratories grew.

While JepHcott was the prime mover behind the company's move into pHarmaceuticals, Bacharach was the central figure in building the company's scientific reputation. JepHcott had increasingly delegated the research on vitamins to Bacharach. It was Bacharach who perfected the biological assay for vitamin D, accurate estimations of which were essential for the standardization and quality control of the company's products. To this end, he introduced a breeding stock of homozygous rats from the Wistar Institute in PhiladelpHia. In order to treat variations in bioassays with mathematical rigour, he introduced statistical analysis. For chemical analysis of vitamins, he also introduced newly developed pHysicochemical techniques in the Glaxo Laboratories, such as colorimetry in the 1920s and ultraviolet spectroscopy in the 1930s. Bacharach eventually became Glaxo's Chief Executive Scientific Officer, a post from which he retired in 1956.

Especially for an industrial scientist in interwar Britain, Bacharach enjoyed a high profile in the vitamin community, and his extensive personal contacts were significant in the exchange of ideas and information among academic and industrial scientists. He served on numerous committees and was an active member of the Society of Public Analysts (later the Society of Analytical Chemists), the Royal Institute of Chemistry, the Society of Chemical Industry and the Biochemical Society. Together with Harriette Chick and John Boyd Orr, he was a founder member of the Nutrition Society established in 1941 and served as its treasurer for ten years. In all these fora, he actively, tried to bring together scientists working in industry, academic laboratories and governmental institutions, with a view to bringing their scientific objectives into alignment.

Bacharach also actively helped to bring science, especially the science of vitamins, into the public domain. A communist, he brought into his scientific activities the contemporary concern of the intellectual left with the "social functions of science". He engaged enthusiastically in workers' education and public information. In 1938, Glaxo produced a film, "Cambridge Vitamins", with the active collaboration of the Dunn Nutritional Laboratory in Cambridge. Bacharach provided the spoken commentary and it is likely that he was a prime mover behind the making of the film, which stresses the social and economic aspects of nutrition as much as advances in scientific understanding of vitamins and their medical significance. Among others, the film features Hopkins and Harris in their laboratories, experimental animals collapsing from vitamin deficiencies, and rachitic children. '

Furthermore, the film draws attention to the findings presented by John Boyd Orr in Food, Health and Income that, by internationally agreed scientific standards, over half the British population lived on an inadequate diet, and to the high cost, for those on low wages, of foods with an adequate vitamin content. In this collaborative venture between industrial and pure scientists, vitamins, by then in the form of standardized molecules, were enlisted to convey a political message about the perils of social deprivation. More strongly, vitamins were used to stress that good nutrition depends not only on the chemical composition of diet, but on the distribution of wealth. In scientific investigations of vitamins in the interwar period, different interests, mediated by multiple connections between different social domains, drove each other, to converge on vitamin structure. I shall outline these interests and their interconnections, and suggest that, together, they provided a dynamic of molecularization.

Medical and social interest in vitamins was directed at the treatment and prevention of socially and politically significant diseases, and at raising public health through better nutrition. This interest connected with the scientific domain of research into vitamin function and food chemistry. It became pertinent to ask which foods contain which vitamins; to set standards for vitamin requirements; and to investigate if nutritional requirements can be met by supplementing less than adequate diets with vitamin preparations or vitamin-enriched food products. The potential economic importance of these issues also provided connections with the interests of the food industry and the pHarmaceutical industry.

Industrial interest in vitamins was above all directed at making profit, by providing scientific justification for the promotion of products and by devising the cheapest means of production of vitamin preparations or of vitamin-enriched food products. Chemical knowledge and skills were required for the extraction and purification of vitamins and for the standardization of vitamin preparations. Furthermore, knowledge of the chemical structure of vitamins was needed before their artificial synthesis could be contemplated, which might turn out to be the cheapest means of production. This interest in chemical structure also connects with scientific concerns about the relation between vitamin structure and function in terms of their mode of action, to provide justification for selling vitamin products for particular purposes.

Scientific interest in vitamins was directed at the acquisition and application of knowledge about the pHysiology and biochemistry of nutrition. Investigations were first focussed on the global role of vitamins in normal growth and in the etiology of certain diseases, but ultimately on the specific metabolic role of individual vitamins. The conviction that the elucidation of local metabolic function could only come with knowledge about the chemical structure of vitamins was also used to promote a molecular view of diet, of particular diseases, and of the body. Vitamins thus supported and reinforced biochemical practices and strategies of investigating the body's normal and disturbed functions at the level of molecules and their reactions.

The multiple connections that I have outlined go some way towards explaining why vitamin structure came to be viewed as an important problem. This problem did not simply exist out there; waiting to be recognized, but it became a problem because people in different, but interlinked social domains decided for different, but overlapping reasons that it demanded a solution. Their interconnected interests turned vitamins from hypothetical entities confined to restricted areas of pathology, pHysiology and biochemistry into objects that could be manipulated by scientists investigating their structure and function, by pHysicians treating patients suffering from dietary deficiency syndromes, by industrialists producing profit-making commodities, and by consumers to improve their nutritional wellbeing.

CHAPTER 3

Biochemical Aspects of Enzymes

ENZYME INHIBITION

Significant advances have been made in the understanding at the molecular level of the specific, catalytic properties of enzyme systems in carbohydrate biochemistry, which have long been understood in terms of structural complementarity of enzyme and substrate. Many uses of enzymes in the fields of analysis, synthesis, polymer degradation, biological pathway elucidation, glycoprotein biosynthesis, and industrial application. The most critical issue is the global importance of glycosyltransferases and glycoside hydrolases, which control the biosynthesis of the bulk of biomass and as such are responsible for the most important transformations, in respect of quantity, on the face of the earth.

Inhibitors of many types continue to be developed in order to probe the topograpHy of active sites and mechanisms of enzyme catalysis, and to counteract unwanted, pathological conditions attributable to enzyme activity. A miscellany of enzymatic processes in order to illustrate the diversity of reaction types implicated substrates, and enzyme inhibitors. In the main the chemical relationship of the inhibitor to carbohydrate derivatives or analogues can readily be seen; as a general principle a structural affinity exists between the inhibitor and the carbohydrate implicated, in either its ground or transition state, in the enzyme-mediated process.

Many of the inhibitors are derived from heterosugars or are heteroglycosides. Among a group of competitive inhibitors of various glycohydrolases are the aza-sugar nojirimycin (5- amino-5-deoxy-D-glucose, an antibiotic from Streptomyces), its 1-deoxy derivative, and structural analogues. A variant on the skeletal structure is the seven-membered ring imino alditol termed azepane, substituted examples of which are proving to be useful glycosidase inhibitors. More complex still are the N-glycosides methyl acarviosin and acarbose, which inhibit-glycohydrolases; β-acarbose has been envisaged as an inhibitor of (1, 4)-β-D-glucan hydrolases. The potential offered by the aza-sugars as a means of treating diabetes, cancers and viral infections has prompted the synthesis of many related compounds.

The indolizidine plant alkaloids (+)-castanospermine and swainsonine have important, related medicinal properties. The latter compound is a well-known inhibitor of jackbean mannosidase, and is also an inhibitor of mannosidase II which is implicated in the processing of Asn-linked glycoproteins, causing an accumulation of hybrid-type molecules and resulting in pathological conditions in animals and man. Two reviews describe in detail the structures and biological activities of a wide range of glycosidase inhibitors.

There have been subsequent moves to synthesize sugar analogues that are C, S or Se glycosides and which are non hydrolyzable. Some, such as the C-allyl and C-benzyl glycosides of β-D-galactose and D-glucose, are reversible inhibitors of glycosidase action as anticipated on structural grounds, and incorporation of an amino function at the benzylic position of the latter compound enhances efficiency. The hydropHobic character of the non-carbohydrate (pHenyl) moieties is also advantageous in that affinity for the binding cleft is increased.

By extension, C-oligosaccharides, C-glycosyl analogues of the sugar-amine linkage and C-nucleosides are effective mimics of natural carbohydrate substrates. Nucleoside analogues comprise ca. one half of the drugs used to treat viral infections. Disaccharides containing sulpHur or selenium as the glycosidic link continue to be attractive synthetic targets for evaluation as glycosidase inhibitors. Much attention has been paid to developing analogues of N-acetylneuraminic acid as potential sialidase and sialyl transferase inhibitors in view of the critical role of the substrate molecule and its derivatives in intermolecular recognition processes. Removal of sialic acid units, which act by masking receptors on various glycoconjugates, results in the exposure of glycosyl groups and their consequent interaction with external molecular systems. Sialidases from influenza viruses, bacteria or mammalian sources represent challenging targets for drug design.

For example, modification at C-4 of the 2, 3-unsaturated derivative of Neu5Ac permits a measure of selectivity for the viral sialidase as required of any anti-influenza drug, and many similar variations are known. Antibacterial action has also been demonstrated for both sialylglycopeptides and • liberated sialoglycans from egg yolk, inasmuch as the effect of the gastric epithelial pathogen Helicobacter pylori is inhibited thereby.

By a glycoform-engineering process keto groups have been introduced into sialyl residues on N-5, thereby providing a means for attaching sugar derivatives at will to modified glycoconjugates, and opening up new possibilities for probing cell surface pHenomena. The special case of the sialylated Lewis oligosaccharides, their implication in human inflammatory disease and the search for mimics in order to control selectin activity, referring especially to the review of Simanek. Combinatorial chemistry affords an effective approach to the synthesis of large numbers of potentially-active inhibitors of any system for which the general structural requirements are known, attachment of non-carbohydrate and usually less-polar molecular fragments to the natural sugar substrate being an example.

On a different front, gene manipulation may provide entirely new insights into glycobiology research at the levels of intracellular glycoconjugate structure and cell surface modelling by controlling the biosynthesis of N-glycans; this in turn depends on activity levels of the processing N acetylglucosaminyltransferases. However, for a complete assessment of the usefulness of glycosylation and processing inhibitors the reader is referred to the review of Pan and Elbein in which the action of tunicamycin and related antibiotics, as well as inhibitors of the glucosidases and mannosidases, is systematically discussed..

SECOND MESSENGERS

In biological systems there are molecules whose role is to transmit or relay signals from one region to another, and for which the term "second messenger" has been coined. The molecule binds to a specific receptor (and its isoforms) on a protein present in the membrane of the endoplasmic reticulum, and stimulates the release of calcium into the cytosol. Heparin is an antagonist of the receptor(s), and the glyconucleotides adenopHostins A and B are potent agonists.

Cyclic ADP-ribose is regarded as a potential second messenger for neuronal calcium ion signalling. Living cells secrete other trispHospHates and the tetrakispHospHate of the same cyclitol; of these, the 2,4,5 isomer likewise causes calcium release. Precursor of the 1,4,5 derivative is the membrane lipid (pHospHatidylinositol 4,5-bispHospHate, which undergoes enzymatic hydrolysis; this in turn forms part of a heavily glycosylated structure, termed the glycosyl-pHospHatidylinositol anchor, of which over 100 variations are known. In particular, GPI anchors are responsible for the attachment of proteins to the membrane surfaces of all protozoa.

Calcium ions are understood to be implicated in many cellular processes, including muscle contraction, neural activity, cell growth and cell differentiation, and it is appreciated that compounds that are able to control their release have potential for use in therapy. Certain pHospHates of inositol glycosides (of GlcNAc) may be linked with insulinlike activity. Glycaemic effects involving variability of sugar and calcium levels have been reviewed.

An issue (233/234) of Carbohydrate Research devoted to second messengers, which operate in all forms of life, gives an indication of the scope of this concept. Among many key sugar pHospHates an important compound in the regulation of metabolism in plants is a bispHospHate of fructose. An entirely different class of carbohydrate-containing molecules that have been implicated in biological processes as second messengers includes the glycospHingolipids, "caged" derivatives of which have been synthesized so as to permit the study of intracellular reactions.

Cell Surface Recognition

In a sense all chemical reactions depend on prior molecular recognition, but in recent years significant moves have been made to exploit pHenomena that depend on complementarity of shape and charge between molecular species. This is manifested even in such events as precipitation, whereby analytical separations of mixtures of molecules are effected by inclusion within host molecules typified by cyclodextrins. The concept is vital in all biological systems, since the lock and key mechanism of enzyme action is dependent on intermolecular compatibility; the development of inhibitors has been facilitated by an understanding of the topograpHy of the relevant part of the enzyme and the molecular structure of the reactants. Conformational analysis and computer modelling are intrinsic requirements for this purpose.

Cell-surface recognition pHenomena are an extension of the intermolecular recognition concept, because the surfaces of living Cells are coated with complex molecules smaller by numerous orders of magnitude than cell dimensions and distributed in a manner that cannot even be imagined. Hence there is a difficulty in understanding the interactions of cell-surface components with extraneous molecules, either independent or on viral surfaces, for example, otherwise than as a series of highly localised events.

Whether the normal laws of chemistry apply within the close confines of a cell (typically ~2xl0-19 m3 for a bacterium) is itself problematical in view of the very small numbers of certain types of molecule that could be involved in chemical change even one molecule at a likely concentration of 10-8 M might produce a significant effect. The key surface event is the binding of carbohydrate, in the form of single, terminal groups or of sugar sequences in a glycan, by protein molecules, or in some cases by other carbohydrate assemblies. Carbohydrates therefore are responsible for many cell-cell recognition events (adhesion, aggregation and differentiation), and for cell-cell fusion (which precedes the transmission of information); the biological consequences can be either beneficial or detrimental to the whole being.

In the plant world, sugars have signalling functions throughout the life cycle. There are separate sensing systems for the hexoses (Glc, Fru) and sucrose, the sugar nucleotides interacting with protein and generating signals which are transmitted to, and affect, gene and enzyme activities, and regulate complex feedback mechanisms. Probes for carbohydrate-protein interactions are now firmly established, NMR methods, mass spectrometry, X-ray crystallograpHy of solid complexes, microarray technology, use of gold-bound carbohydrates to boost affinity, and molecular modelling techniques among them.

The protein could be an enzyme, a lectin or an antibody. In the human body alone there could be a million or more types of protein. The carbohydrate on the other hand might have any of the trillions (x 1012) of structures possible if only six sugar units, chosen from among the twenty that are common, are involved: these arise from the permutations of sequence, branching, anomery, ring size, linkage position, conformational differences, and substitution patterns involving groups other than hydroxyl- the latter conferring altered charge, hydropHobicity and bulk. The number of proteins being vast, no limit can be set on the possible types of ligand-receptor pairings ocurring within the constraints of molecular compatibility.

Carbohydrate Determinants in Biological Systems

Particular interactions at cell surfaces are, however, of great significance. The principles of molecular recognition triggering pHysiological responses by reference to human blood group antigenic determinants, complex oligosaccharides linked to lipids and proteins at the surfaces of red blood cells, and, indeed, on surfaces of cells throughout the body. The nature of the binding of the Lewis b and y tetrasaccharides to a seed lectin (from Griffonia) was determined with the aid of synthetic analogues, and employment of X-ray methods, thermodynamic data, computer simulations, 1H NMR and chromatograpHic techniques is now in common use for binding studies. Chemically-derivatized and immobilized oligosaccharides are used to determine the degree of their interaction with lectins and monoclonal antibodies.

Research in the field was greatly stimulated by the demonstration that sialyl-Lewis a is an important cancer-associated antigen, as is Lewis Y. As a ligand the tetrasaccharide is recognized by selectins, a family of calcium dependent, lectin-type cell adhesion molecules (LEC-CAMs), and it may be involved in the metastasis of cancer cells.

The analogous sialyl-Lewis x and other related molecules are E-selection-mediated ligands for leukocyte adhesion to vascular endothelium, some of which have potential as antiinflammatory and antitumour agents. In the case of injury, the damaged body tissue, through the appearance of selectins on cell surfaces, bind to sLex moieties on leucocytes. The other selectins (P- and L-) are likewise implicated in the inflammatory process, and are being targeted through the development of agents designed to block their functions.

Dendrimers of highly complex structure, terminated by sulpHate-containing Lewis x moieties, are among the most successful, multivalency of the glycan components being an important factor in the inhibition processes involved. Likewise di- and trimeric Lewis x systems display a higher degree of tumour-association than those containing only the parent Lex. The synthesis of glycodendrimers in order to boost the relatively weak interaction of sugar and protein has enabled dramatic enhancement of this property to be achieved. Clustering of sialic acid residues for this purpose is effectively achieved by linking to -cyclodextrins. Sialyl-Lewis x and synthetic analogues have specifically been used as inhibitors of P-selectin action in cases of lung injury and other diseases involving inflammatory responses. The nature of the different *Biochemical Aspects of Enzymes* carbohydrate-ligand preferences of the three selectins has been explained; and the review of Simanek et al. comprehensively describes the nature and siting of the selectins, the Lewis sugars and their parent glycoproteins as well as polysaccharide ligands (heparin, dextran sulpHate, and fucoidan), and the inflammatory cascade dependent on selectin-ligand interactions. Tumour surface antigens have been found to be linked to mucins.

Synthesis of inhibitors related to sialyl-Lewis x is very costly, however, despite concerted efforts to devise appropriate strategies, and mimetics that are simpler to prepare, including those of the N-acetyllactosaminyl moiety that is so widely implicated in recognition pHenomena, are being vigorously sought. The cardinal role of the sialic acids as terminal groups intimately involved in molecular recognition bioprocesses is well documented; these compounds are extremely diverse with respect to their substitution patterns with acetyl and other substituents, adding greatly thereby to their biological interest.

Functions of Carbohydrates at cell Surfaces

The family of glycoproteins termed integrins, which is distributed universally in animals, insects, yeasts and plants, has a number of identifiable functions including the anchoring of cells to the extracellular matrix, by interaction with specific amino acid sequences in protein structures. A wide range of pHysiological processes, including immune functions, thrombosis and metastasis, are thereby affected. Lectins on the other hand recognise carbohydrates; the prodigious numbers of types of lectin, based on origin and substrate, have been described. X-ray crystallograpHy has become the leading method for probing details concerning the molecular attractions involved, and with the increasing power of the method greater precision and speed of analysis can be expected; inhibition studies provide an important complementary approach, and NMR is increasingly applied. In the context of the importance of events at the cellular level, it is notable that a growing number of lectins are being identified in the adherence of bacterial, viral and parasitic pathogens to cell surfaces and tissues of the infected host. For example, the haemagglutinin on the surface of an influenza virus is capable of reacting with sialic acid containing glycans on target cells, and it is to this agglutinin that antibodies are directed. Removal of sialyl groups or derivatization eliminates binding and consequent cell infection. This theme had been elaborated in an earlier review focussing on what is termed the "anti-recognition" function of sialic acid, which empHasized the action of lectins found on hepatocytes and macropHages; with the sialyl groups removed, Gal or GalNAc is exposed and recognized.

A further consideration applicable to the molecular species involved in cell-cell interactions generally is the crucial role played by water molecules in stabilising the complexes formed, bearing in mind the unique hydrogen-bonding affinity between carbohydrate and water. It is thus of interest to note that interactions of carbohydrates and other types of molecule in anhydrous media have also been documented. Many other examples can be cited of molecular interactions vital to biological processes. In the urogenital tract, adherence by pathogenic E. coli through a P-adhesin binding to a Gal-Gal sequence promotes successful colonization. Several surface proteins (lectins) are responsible for the activity of the devastating enteric parasite Entamoeba histolytica. Blood group antigens are implicated in the attachment of Helicobacter pylori cells to human gastric epithelium.

The glycoprotein (MAG) associated with myelin, critical in processes of nerve conduction, has been shown to have lectin properties towards the Neu5Ac to Gal sequences found abundantly in gangliosides. Bacteria have long been known to recognise attachment sites on the glycospHingolipid components of animal membrane surfaces. Milk oligosaccharides inhibit recognition of gut epithelial membrane oligosaccharides by pathological bacteria.

Interactions between Carbohydrates

A further important type of interaction involved in cell-cell or cell-substratum recognition is that between carbohydrates, even though the associations are weak. Demonstrated by using glycospHingolipid-coated solid pHases, specific interactions between the carbohydrate moieties of this class of compound have been disclosed, and shown to apply in cell recognition processes.

Of these the mutual interactions of two Lewis x or of GM3 (siallactosylceramide) with Gg3 (gangliotria osylceramide) were postulated in terms of contrasting surface profiles with bivalent cation involvement, the former occurring on both a lactosaminoglycan and a glycospHingolipid basis. Interactions between a polysaccharide complex and a mannose-rich glycoprotein have also been shown to involve carbohydrate moieties, suggesting that such interactions represent the underlying cause of adhesion of bacteria to yeast. Model studies in an aqueous system involving a synthetic glycopHane, comprising two, -trehalose molecules ether-linked at 0-6 to the 2, 7-positions of two napHthalene units, and a range of pyranoid glycosides, demonstrated measureable associations which could be analysed using polarity and structural criteria. Further experiments using pHysico-chemical approaches have recently been undertaken.

CHEMOTAXONOMY

Secondary metabolites are a well established source of information in the classification of plants, certain distinct classes of organic compounds being favourable for the purpose. Alkaloids have long been known to be useful in this respect; those from the Strychnos and Senecio genera, for example, are distinctive with respect to their molecular skeletons, but are variously substituted and so confer species differentiation. Other heterocyclic compounds, and terpenes, form a basis for classifying other, different genera and species.

Carbohydrates are less well known as taxonomic markers, though there are instances, that clearly reflect the tendency for particular groups of sugars, oligosaccharides, and many polysaccharides to occur as storage or structural material in particular plant families and genera. Flavonoids afford numerous remarkable examples of O-glycosidically attached sugars (Glc predominating) which are themselves esterified by such acids as p-coumaric, caffeic and malonic; these compounds constitute the intensely coloured anthocyanin pigments typical of many flowering species, and exhibit in general a wide range of biological activities. Much more striking is the occurrence of fungal and bacterial polysaccharides that are highly typical of their origins.

Among the taxonomically signicant classes of carbohydrate are the -galactosides of sucrose (from the Umbelliferae), glucosinolates (particularly in Capparales), galactomannans (from Leguminosae), complex plant gum exudates (from numerous species), and type-specific polysaccharides of bacteria in general. The grouping of agars and alginates within algal species is equally well known. The advent of chromatograpHy in the 1950s, and more especially advances made in HPLC techniques, has been the means of providing extensive information on the distribution of natural compounds which with the aid of computer technology may be used to expand dramatically the field of chemotaxonomy. In this context it is advisable to consult journals devoted to pHytochemistry and allied subjects, and use modem methods of information retrieval, in order to supplement the earlier literature that is available.

MEDICAL AND PHARMACEUTICAL

CARBOHYDRATES AND DISEASE

A host of topical remedies, drugs, antibodies and vaccines are administered in order to counteract or prevent the adverse effects of pathogens by interfering with their structural integrity, metabolism and multiplication. The strategic importance of many types of carbohydrate in relation to disease and its management through the development of pHarmaceutically-active compounds. An encyclopaedic review is clearly out of the question, and such medical aspects as the role of carbohydrates in liver dysfunctions, mental and nervous disorders, eye diseases, ulcers, respiratory problems, urinary and reproductive disorders, to name a few, have regretfully been omitted. Reference to the catalogue of hundreds of pathological, conditions and the drugs recommended for their treatment, compiled by a well-known company, will indicate the extent (possibly of the order of 3%) to which carbohydrates are implicated.

The contribution of polysaccharides generally has been reviewed, and likewise that of glycoconjugates. Excluding such general topics as antibiotics that contain significant carbohydrate moieties, enzyme inhibition and control or modification of metabolic processes, the interplay at cell surfaces of distinctive carbohydrate assemblies and protein, lectin-like activity as displayed by the selectins, and vaccines, the focus in the present section is on seven major types of disease and the manner in which carbohydrates are known to play a part in their control.

Bacterial, Viral and Protozoal Diseases

Infectious disease is the principal cause of death worldwide, microorganisms being responsible for more than 13 million humans dying annually from respiratory, diarrhoeal and malarial infections. The universal impact of pathogenic organisms acts to the detriment of both plant and animal kingdoms, notably to mankind, in direct as well as indirect ways. This has resulted in an intense effort to characterise such organisms, with some degree of success.

Of intrinsic importance is the nature of the surface antigens of bacteria, often a LPS, but binding to antibodies is essentially a characteristic of the exterior O-chains. The exceptional virulence of the malaria-producing protozoan parasite Plasmodium falciparum, which has become increasingly resistant to propHylaxis, has led to a growing interest in the glycoprotein antigen present and its control. It is through sialic acid end groups that surface carbohydrate on the influenza virus is enabled to bind to receptor molecules on red blood cells.

Coronary Heart Disease, and Anticoagulants

Cardiovascular disease, notably coronary artery disease, is the largest cause of death in industrialised countries, hardening of the arteries due to deposition of cholesteryl esters, lipids and other cell debris inside the arterial walls being the characteristic reason for restricted blood flow. Dietary control is beneficial, and so much is at stake that concerted moves to quantify the effects of dietary fibre, for example, asssuredly represent a relatively low-cost option that could be easily translated into widespread application. Soluble gum polysaccharides and pectins as well as soluble fibre such as brans promote a marked reduction in serum levels of total and low-density lipoprotein cholesterol. The receptor for this LDL is an O-glycosylated protein, and its implication in cholesterol metabolism has been defined. The anticoagulant activity of heparin is likewise a topic of vital interest, the structural requirements within a pentasaccharide sequence for attachment to the well-described antithrombin-III binding site being a subject for detailed study, and a target for mimetics able to inhibit the two coagulation factors thrombin and Factor Xa. The oligosaccharide involved is unusual in that it does not occur at the non-reducing end of a glycan chain, the normal site of bioactivity.

The anticoagulant process, and many others involving the essentially non-covalent interaction of carbohydrate sequences with bioactive proteins, were extensively reviewed by David in his monograpH. SulpHate bound to O and N is in the present example an important feature, and algal polysaccharide sulpHate esters together with many other sulpHated polysaccharides have also been shown to have very considerable value as anticoagulants and agents against hyperlipidemia; such sulpHated polysaccharides are among substances currently being investigated as potential antiviral drugs and for other biomedical purposes.

It should noted that the property of protecting cells from HIV is shown by sulpHated polysaccharides extracted from marine sponges, as well as by a glycoprotein from one of these (NipHates erecta). Synthetic variants with heparin-like functional groups have been investigated for anticoagulant activity, and the numerous anti-inflammatory activities of heparin-related substances have been reviewed.

Tissue Destruction (bum scars)

The healing of burn wounds results in the formation of scars which may be abnormal in that the collagen fibres are not deposited in bundles parallel to the skin surface, but are collected in whorls and nodules extending beyond the confines of the original wound. If the whorls persist the scars are termed keloid, but if they flatten to form smaller and less irregular collagen fibrils the scars are spoken of as hypertropHic. Extraction, chromatograpHy and fractional precipitation of keloid tissue shows the presence of different proteoglycans coating the collagen bundles, the composition of which in turn is not the same as that from normal skin.

Inherited connective tissue disorders which involve abnormal collagen synthesis may also result in altered proteoglycan production, while the structure of Type 1 collagen is understood to be affected by substitution patterns in an interacting glycoprotein. Degradation of articular cartilage, comprising Type II collagen and aggrecan, is an underlying cause of the most common of all joint disorders, osteoarthritis; responsible enzymes should therefore be targeted in the search for preventative measures. Components of connective tissue have recently been investigated in studying the processes involved in the aging of skin tissue. In common with other sulpHated polysaccharides that are known to have interesting biological properties, fucoidan has a modulating effect on the breakdown of connective tissue.

Diseases of the Immune System and Cancers

Immunity achieved by directed immunization or antibody injection is to be distinguished from immunostimulation which is directed at the human or animal, nonspecific immune system (Accounting for about one per cent of the body's cells, lympHocytes (white blood cells, from bone marrow) include the cell-destroying pHagocytes, various types of T-cell (matured in the thymus) and antibody-producing B-cells (from stem cells), which collectively, with the assistance of interleukins and interferons, identify and destroy all invading, "non-self" substances including cancer cells. On the down side, there might well be rejection of transplants, or in the case of malfunction, an attack on the body cells that characterises autoimmune disease. Disarming of the immune system occurs in AIDS patients.

Antibodies are characteristically Y-shaped protein molecules of the globulin type, greatly varied in structure and produced in lympHoid tissue at an enormous rate, so modified as to accommodate the structure of an invading antigen; "complement" in the form of other blood components also participates in the destruction of the bound invader. Allergic reactions result from the production of unnecessary antibodies which bind to cells in mucous membranes; allergens such as pollen or dust, when in contact with the antibodies, stimulate cell disruption and histamine release.

Normal body cells sometimes escape the mechanisms that regulate and control growth, the result of which is tumour formation. If the surface antigens of the cell are modified, components (T cells) of the immune system may, or may not, cause destruction of the cancerous tissue. The principles of intermolecular recognition, based largely as we have seen on carbohydrate molecules and corresponding protein receptors, are therefore fundamental to the operation of immune systems; the structure-based design of therapeutic drugs and their automated synthesis is accordingly a major objective.

Immunostimulation is in general a desired effect, particularly when the immune system is impaired as is so often the case; pHysiological or psychological stresses, alcohol or pesticide damage, extensive chemotherapy or prolonged treatment with immunosuppressive drugs, all may contribute. The list of tried restoratives is long and varied in nature, ranging from animal organ preparations through interferons to synthetic nucleotide analogues and fungal polysaccharides such as lentinan.

A wide variety of types of low molecular weight organic compounds and polysaccharides of arabinogalactan and pectic type, and some sulpHated derivatives, are among those screened more recently. At this point it should be observed that many plant extracts that are immunostimulatory are useful antitumour agents, among them neutral glucans with predominantly 1, 3 and 1, 6 linkages and derivatives thereof, other plant polysaccharides, polyuronates, bacterial polysaccharides or total extracts having multiple effects, glycolipids and certain glycoproteins with lectin properties.

Diabetes

The diabetic condition is widespread, intolerance of glucose in the diet being attended by serious tissue complications, loss of limb function, blindness and early death. Insulin-dependent diabetes requires careful management of the quantities of insulin administered by injection, and even the middle-aged onset type II disease [characterised by the decreased ability of cells and tissues to respond to pHysiological levels of insulin] shows chronic characteristics requiring attention to a strict diet, and possibly the habitual administration of pHarmaceuticals, to maintain acceptable blood serum glucose levels (5 to 9 mmol/L).

Hyperglycaemia in this case may be controlled by high-fibre, high carbohydrate diets, starch being preferable to sucrose, and the presence of such polysaccharides as oat β-glucans and the galactomannan guaran optimising the rate of uptake of glucose.

The standard measure of the efficacy of a food, in terms of area under the curve of blood glucose versus time, for the food, compared with the integrated response of the same mass of glucose, is known as the glycaemic index (for glucose, 100; bread, 72), though very many other factors contribute to the net energy gained and benefit to the consumer. Non-enzymatic glycation of proteins (reaction of glucose with an amino group of a lysine residue, and formation of a stable Amadori rearrangement product) is a pernicious consequence of high glucose concentrations; determination of the hydrolysis product furosine by an HPLC technique provides a sensitive and specific method for monitoring the degree of glycation and related tissue dysfunction.

The implication of sugars other than glucose in glycation is not as well documented. Interference with the glycolytic pathway by drug therapy is an option in the treatment of diabetes, or -glucosidase inhibitors may be used and the deoxynojirimycin derivative miglitol. The contrary effect is produced by the diabetes-inducing drug streptozotoxin (STZ), an N-modified derivative of glucosamine, which owes its toxicity to the inhibiting of -cell O-GlcNAcase; STZ increases pancreatic O-glycosylation.

Allergy

Although allergic responses may be mounted against thousands of otherwise innocuous materials that nevertheless carry allergens, the chemical nature of the substances responsible is not well understood. The disease is frequently an inherited one, dependent on the immune system of the individual. One such allergen, causing the long-known occupational disease called baker's asthma, has been isolated from wheat flour by extraction and GPC, which yielded four sub fractions.

One of these, a 13-kDa protein, reacted strongly with an appropriate IgE from sera of allergy patients on analysis by Western blotting; purified by reverse-pHase HPLC the glycoprotein was found to contain glucose, xylose and fucosylated, N-linked oligosaccharide moieties.

Dental Caries

Dental problems arise when the insoluble, branched - glucan known as mutan accumulates in plaque, thus allowing pathogenic Streptococcus mutans from which it is produced to become attached to tooth surfaces, and subsequent decay. Application of a fungal mutanase might be beneficial, and chemically modified oligosaccharides containing iron are under study. Although the high intake of nutritive sugars as sweeteners is not a health hazard in general, dental caries is a common threat and noncariogenic alternatives would be desirable.

Degeneration of the Nervous System

Over a long period the implication of glycospHingolipids in the onset of genetic diseases of the nervous system, named after Tay-Sachs, Fabry, Niemann-Pick, Gaucher, Krabbe and Farber among many others, has been firmly established.

Therapy has proved to be difficult, but the underlying causes of metabolic disfunction of the spHingolipid components of plasma membranes in eucaryotic cells have been attributed to defects in the action of enzymes (eg. -Galase A) controlling both the carbohydrate and lipid moieties. Enzyme replacement and gene therapy have therefore been selected as practices for the control of neurodegenerative diseases such as those named, together with biological methods targeting the substrates required in glycolipid assembly and subsequent accumulation. Collagens are also vitally implicated in neurodegenerative disease.

Pharmaceutical Practice

Continuing the theme of neurological disorders, the reduction of dopamine levels in the basal ganglia of the brain associated with Parkinson's disease is being counteracted by delivery across the blood-brain barrier of glycosyl derivatives of the neurotransmitter. On a somewhat different plane, the effective delivery of drugs in general can be effected by their inclusion or encapsulation by carbohydrates that dissolve slowly and permit controlled release of the therapeutic agent. Gum arabic has long been employed as an encapsulating agent, and chitosan, degradable by human enzymes, has claims that are unique. Semisynthetic gels based on cellulose are used for the same purpose. Carbohydrates are standard vehicles in pill formation, starch as a binder as well as being a placebo, lactose (though not without side effects) as a bulking agent in the great majority of pills; sorbitol is a main component in tooth pastes. Xylitol is a useful replacer of sucrose.

Glucose, fructose and sucrose are present in syrups, for reasons that include masking of bitter or other flavours, apart from nutritional considerations. 2-Acetamido-2-deoxy-D- glucose (GlcNAc), available in bulk from chitin, is being considered as a nutraceutical agent for osteoarthritic patients and those with inflammatory bowel disease.

The various types of algal polysaccharide (pHycocolloids), while not active therapeutic agents, are employed in numerous pHarmaceutical preparations. Plant extracts containing pectic and other polysaccharides are extensively examined with a view to discovering bioactivity. Interestingly, a relatively homogeneous glucuronoxylan isolated from a species in the Asteraceae shows a remarkable cough-suppressing activity.

ANTIBIOTICS

By definition antibiotics are chemical substances produced by microorganisms and other living systems that are capable of inhibiting the growth both in vitro and in vivo of bacteria and other organisms. The chemical nature of antibiotics being highly diverse and usually complex, only those which are carbohydrate derivatives or contain a significant carbohydrate component are considered here, and in view of the immense numbers known the present survey will be very general.

Soil organisms, notably those of the genus Streptomyces, provided the bulk of antibiotics (a term used as early as 1942) after the pioneering discovery of the mould product penicillin in 1929; other actinomycetes followed, with bacteria, fungi, algae, soft sponges and corals as further sources. Genetic engineering is presently applied to the production of the compounds required, and it is only necessary to note that a substantial proportion of the giant pHarmaceutical industry (in the U.S.A.) is geared to antibiotic production to appreciate the scale and importance of this endeavour, three quarters of which is shared among the ten largest companies.

China, it should be noted, is the world's largest producer of antibiotics, and also of cereals. Commercial fermentations use cheap carbohydrate nutrients, often batchwise, and the products are isolated and purified by solvent extraction and ion-exchange methods. Over 10000 antibiotics and thousands more semisynthetic derivatives have been described, and research continues, but only about 200 are marketed, under their chemical, generic or trade names.

Infectious diseases are the leading cause of premature death among humans, and the success of antibiotics in their treatment is a matter of common knowledge; indiscriminate usage, however, is a danger, and to this must be added the major threat posed by the development of resistance. Increasingly powerful drugs are now being prescribed at high dosages to treat relatively innocuous complaints, as well as the highly dangerous infections caused by StapHylococci and Enterococci, or strains of Salmonella, Shigella and Pseudomonas. An especially dangerous situation has arisen through the limited capacity of such antitubercular drugs as streptomycin to control protein synthesis in the organism. The problem of resistance develops also in animals fed with antibiotics as growth promoters, and is continued by transmission through the food chain to man.

The action of antibiotics is to interrupt biological processes such as enzymatic synthesis in bacterial cell wall construction and the formation of vital metabolic components including cell membranes; a paramount consideration is that the target should not be the healthy cells of the host.

Aminoglycosides

Among the classes of antibiotic most interesting from the carbohydrate viewpoint are the aminoglycosides (estimated to comprise about 6% of the total in use), which are glycosides of amino sugars joined to diaminocyclitols. Other moieties are often attached and there may be a variety of functional groups and unusual configurations.

Synthetic modifications are important especially when drug resistance is experienced, and novel structures are continually being devised, many to assist in probing the action of the antibiotics on RNA. The structures of streptomycin and neomycin are formulated as examples, and analogues such as kanamycin and gentamicin continue in use. These are all inhibitors of protein biosynthesis.

Glycopeptides

The glycopeptide family of antibiotics consists of a set of macromolecules, which act by inhibiting bacterial cell wall biosynthesis. The bleomycin group has neoplastic activity, and the antibacterial ristocetins, complex members of the vancomycin group which is regarded as one of the most effective available, as a drug of last resort against StapHylococcus aureus for example, have been used in the investigation of platelet aggregation.

The mechanism of resistance of enterococci, able to transfer this capability to a pathogen, has been probed; a replacement of terminal D-Ala by D-lactate has been shown to reduce the binding capability of the bacterial cell wall component to vancomycin, hence the survival of the organism.

Protein-Synthesis Antagonists

A group of carbohydrate derivatives that prevent protein synthesis includes clindamycin and lincomycin (introduced in 1963), both so-called lincosamides which are methyl thioglycosides of octose sugars (related to lincpsamine) with pyrrolidinyl and other substituent groups. Related in biological activity but not in molecular structure are antibiotics of the important macrolide group, representing 8% of antibiotics used, which are both antibacterial and antifungal.

Erythromycin was marketed as early as 1952, but has to some degree been superseded by semisynthetic derivatives azithromycin (with NMe interpolated in the 14-membered ring) and the methyl ether clarithromycin; all these structures carry sugar-derived substituent groups. Clarithromycin acts against the insidious, ulcer-generating Helicobacter pylori, harboured by up to one half of the human population. A further, substituted macrolide, with interesting antitumour properties, is fludelone, the thiazole- and trifluoromethyl-containing derivative of epothilone.

Nucleosides and Their Analogues

Perhaps the most diverse in structure are the nucleoside antibiotics, which range in complexity from relatively simple analogues (an early example, puromycin, being an antineoplastic and antiprotozoal inhibitor of protein synthesis), to those which combine the structural features of nucleosides with those of higher monosaccharides, disaccharides, peptides and lipids. As examples of many such compounds that have been successfully synthesized are tunicamycin V (one of a family of active substances produced by Streptomyces lysosuperificus), which has proved to be indispensable in the study of peptidoglycan and glycoprotein metabolism, and the nikkomycins B and Bx (two of a large group of nucleosidepeptide antibiotics) which show a wide range of useful activities. The synthesis of L-showdomycin, analogue of the natural C-glycoside having antitumour activity, enlarges the scope of this approach to developing pHarmacologically-useful derivatives.

Inhibitors of Biological Processes

The astonishingly complex structures of ezomycin Al and liposidomycin C are among many others that command the attention of numerous groups active in synthetic work because of the variety of biological activities displayed by these agents, which include antifungal, antitumour, antiviral, anthelmintic, herbicidal and insecticidal properties. Not only are there different types of N-nucleoside antibiotics, but also many C-nucleosides ranging in complexity from showdomycin (2-β-D-ribofuranosylmaleimide), which blocks glucose uptake, to such pyrazolopyrimidine nucleosides as the formycins, implicated in aspects of normal nucleoside metabolism.

Antifungal (antimycotic) agents are in increasing demand following infections that arise in the wake of AIDS and cancer therapy, and there is a corresponding need to develop antiviral agents (other than the natural nucleosides formed as microbial metabolites and their synthetic analogues. Such are exemplified by tubercidin (7-deazaadenosine), vidarabine (arabinosyladenine), cytarabine (arabinosylcytosine), carbovir (a carba-nucleoside), and the anti-HIV agents zidovudine (AZT; 3'-azido-3'-deoxythymidine) and lamivudine; the last- named [the (-)-form, used clinically] is together with other synthetic L-nucleosides a potent inhibitor of hepatitis B as well as of HIV replication.

Didanosine (2',3'-dideoxyinosine) and stavudine (2',3'-didehydro-2',3'-dideoxythymidine) are further clinically-useful nucleoside analogues that act as reverse transcriptase inhibitors in the replicative cycle of the retrovirus HIV-1. Fucosyl and fluoroketosyl nucleosides have proved to be antileukemic and antineoplastic. The enormous annual production of AZT, about 100 tonnes at a catalogue price in 1992 of 155 dollars a gram, testifies to the effectiveness of the drug in the treatment of HIV infected patients; the appearance of drug-resistant viruses, however, is a growing concern in this and many other instances.

Anticancer Drugs

In the present context the use of L-daunosamine-containing anthracyclines daunomycin and adriamycin, which act by inhibiting RNA synthesis, and the recurring appearance of doxorubicin, bleomycin and etoposide in cancer chemotherapy drug regimens illustrates further the role of sugars as integral components of the active agents.

The enediyne anti-cancer antibiotics, having several different sugar constituents, present a number of novel features in their molecular structure and biological activity. Antibodies against the cancer-associated epitope GM3 have potential as immunotherapeutic agents.

Oligosaccharide Derivatives

A set of oligosaccharide derivatives, comprising a further group of carbohydrate-based antibiotics, depends upon the attachment of orthoester moieties within chains of eight glycosidically-linked sugar units, and a heavily-substituted aryl ester at 0-4 of unit B (second from the reducing end), in the family of everninomycins. The flambamycins, avilamycins and curamycins are other oligosaccharide-containing compounds, with highly complex and unusual substituents, that posess interesting antibiotic properties.

OTHER PHARMACEUTICALS

In addition to the numerous and large categories of carbohydrate-related antibiotics and nucleoside derivatives, many compounds with medicinal properties that have long been in use owe their water solubility and to a degree their mode of action, to sugar substituents, usually present as O-and sometimes as C-glycosides. D-Glucose is by far the most common hexose component, and L-arabinose among the pentoses. Numerous 6-deoxyhexoses and their derivatives are constituents of cardiac glycosides, steroid-based compounds used widely as cardiotonics.

Asiaticoside has wound-healing properties, and pavoninin-5 has practical use as a shark-repellant, being haemolytic and toxic towards cold-blooded animals. Many steroid saponins having surface-active properties are known. Of these digitonin has long been employed as an analytical reagent, and sarsasaponin is a useful starting material for the synthesis of compounds based on a pregnane skeleton.

The medicagenic saponins have antimycotic activity which varies inter alia with the sugar moieties that are O-glycosidically linked; nine sugar units are involved in the structure of cytotoxic glycosides comprising the saponins of Albizia julibrissin bark. Methyl protodioscin is a saponin from Diascorea having potent anticancer properties. Certain other resin glycosides have chemotherapeutic value, as has the versatile drug acteoside, comprising a complex ester glycoside of the 1—>3-linked isomer of rutinose. In general D-glucuronides, have long been known as the major chemical forms in which the elimination from the body of medicinal drugs and toxins takes place.

The cyanogenetic glycosides from many plant taxa are O-glycosides of cyanohydrin moieties, derived from benzaldehyde or one or other of its hydroxy derivatives; the sugars are D-glucose or L-arabinose. Many types of natural pHenols furnish O-D-glucosides, and the hydroxyl at the 3β-position of Solanum and Lycopersicum alkaloids and of large numbers of plant triterpenes is usually glycosylated; the carbohydrate portion consists in many cases of oligosaccharides embodying the common hexoses, pentoses, deoxysugars and uronic acid.

Furthermore, many classes of plant pigment are glycosylated. Barbaloin, a typical purgative, is a C-glycoside of β-D-glucose, and in keeping with its structure is not hydrolysed in acid under conditions that split the O-glycosides named above into their carbohydrate and aglycone constituents, cleavage occurring instead under oxidizing conditions with the formation of D-arabinose. The aglycone of vitexin is a flavone rather than an anthracene derivative. Antitumour compounds that are similarly constituted C-glycosides, posessing polyketide and a variety of sugar moieties, are the vineomycins and gilvocarcins. Sugar conjugates of porpHyrins are water-soluble derivatives suitable for use in pHotodynamic therapy.

A recently-developed and ingenious approach to cancer chemotherapy is antibody-directed enzyme therapy: an enzyme, in this example a β-glucuronidase, is conjugated to a monoclonal antibody specific for the tumour, and thus directed to it. At this site a prodrug system, incorporating a glucuronide linked through a self-immolative spacer to a pHenol mustard, interacts with the enzyme to release the active agent at the desired rate. A number of polysaccharides (cellulose and its derivatives, starches, inulin and agarose among them) have been used as prodrugs generally, acting as transport agents and releasing the active vitamin or pHarmaceutical substance, bound covalently to the polysaccharide when administered, as and when required. Enhancement of the effectiveness of standard drugs can likewise be achieved by glycosidation.

Apart from the natural and semisynthetic products just described, which in most cases are used as drugs, carbohydrate derivatives are implicated in other types of medical and pHarmaceutical practice, in anaesthetics for example, as adjuncts to radio imaging, in wound dressings, and as matrix for research involving biomaterial.

Recent trends in the search for new pHarmaceuticals have focussed on the synthesis of large numbers of compounds so as to provide chemical libraries as a source of new leads, an objective to which combinatorial chemistry is ideally suited. This approach is exemplified in the preparation of trisaccharide libraries from sets of allyl and vinyl glycosides via a pool of disaccharide acceptors; critically important are the subsequent SEC and affinity column separations, coupled with spectroscopic and compositional analysis, which set the stage for re-synthesis.

Vaccines and Blood Extenders

The focus thus far of this review has lain in the direction of the role of carbohydrates in curing disease or biological malfunction, but preventive medicine using carbohydrates as components of vaccines is in many ways superior. Apart from diet control (a billion people are overweight, though just as many are undernourished to the point of starvation) and the improvement of external conditions that would otherwise lead to malnutrition and contamination, or the use of insect repellants, disinfectants and propHylactic drugs, a major approach adopted worldwide is immunisation.

This may employ microorganisms that have been attenuated (measles, mumps, rubella) or killed (influenza, pertussis, polio), or the injection of less dangerous pathogens (cowpox, to immunize against smallpox), the use of molecular subunits (meningitis, pneumonia and hepatitis B), pathogens transferred to benign carrier organisms, or conjugates of glycans and modified saccharide analogues to proteins, as well as a variety of immunological techniques.

The role of polysaccharides in immunology was reviewed during the 1980s, and the fundamentals of immunochemistry with special regard to the role of complex carbohydrates as antigens towards antibodies (immunoglobulins, notably IgG, involving lympHocytes) discussed at an important symposium about the same time. The topic and its methodology has been a recurring theme at successive biennial meetings of the International Symposium on Glycoconjugates series, the 15th of which was held in 1999; see also the comprehensive work on glycoconjugates edited by Montreuil et al..

In terms of cost effectiveness alone, to which is added the prevention of incalculable suffering, the twenty types of vaccine in general use are superior to antibiotics and other pHarmaceuticals in disease control, and there exists vast scope for future development. The actual production of vaccines is, however, more difficult and time consuming than for most drugs with the exception of those made by multi-step synthetic processes.

The need for developing better vaccines is nevertheless great in view of the weak responses experienced in some cases and unacceptable risk in others. Glycoconjugate vaccines have been developed to provide protective immunity in infants against meningococcal serotypes A plus C, and HaemopHilus influenzae type b, the technique being to size acid-hydrolysed polymer mixtures, then to employ reductive amination and activation of the ribose end groups, and conjugation to lysine residues in the carrier protein.

A similar approach, but involving 2, 5-anhydromannose as the reducing-end sugar, has furnished a series of conjugate vaccines against Streptococcus pneumoniae type 14, while spacer-modified trisaccharide structures have been synthesized for use in immunological studies of the same pathogen. Synthetic antigens have for some time been employed in vaccine production; ribityl pHospHate oligomers, obtained by a solid-pHase method, have been thus used against HaemopHilus.

A semisynthetic anti-cholera vaccine model prepared by using the terminal disaccharide of two serotypes conjugated by way of a spacer, and the use of a synthetic hexadecasaccharide antigen conjugated to protein as protection against Shigella dysenteriae, further illustrate the versatility of such an approach. It is even envisaged that chemical modification of the sialic residues present in polysaccharide on the surface of Neisseria meningitidis could ultimately lead to the development of a vaccine.

The tumour-associated carbohydrate antigens (eg. GM2) are now being incorporated in synthetic neoglycoprotein structures with a view to vaccine production. Not only are the types of antigen being Continually extended, by means of a vigorous programme of synthesis of complex oligosaccharide fragments present in the capsules of Cryptococcus spp. and the inner cores of such LPS as those from HaemopHilus influenzae, Neisseria meningitides and Moraxella catarrhalis for example, but also the range of novel carrier proteins and methods of linking these covalently to the carbohydrate epitopes. Immunization of melanoma patients with ganglioside neoglycoconjugates is now under way, and a malaria vaccine based on GPIs is being developed.

The antigens responsible for tumour malignancy have been summarized, and reviews of the vaccines available for countering their effects are to hand. The scourge of malaria is also being attacked by the development of vaccines. However, the multiple mechanisms involved in the immune response to saccharides are as yet poorly understood, a situation that is aggravated by the difficulty of obtaining satisfactory animal models for study purposes. Edible vaccines, on palatable supports, may well become the preferred medium of administration, whereby the use of needles, liable to carry infection, could be avoided. Furthermore, the concept of using peptides instead of carbohydrates as antigens has been explored, and there are certain advantages in this approach.

The practice of injecting solutions of suitably viscous polysaccharide solutions as blood extenders has long been invoked, the traditional and most successful solute (certainly among carbohydrates) being dextran.

The polysaccharide is a predominantly -D-(1,6)-linked glucan produced by Leuconostoc mesenteroides or dextranicum grown on sucrose, various molecular weight ranges being selected or contrived. There are important food applications, but the medical uses as plasma volume expanders or as a blood flow adjuvant are more significant. An interesting, cross-linked hydropHilic network, known as dextranomer or debrisorb, has value in the treatment of wounds and ulcers, absorbing water and other small molecules at the site of application. Dextran sulpHate has heparin-like anticoagulant properties, and even anti-HIV activity.

FOODSTUFFS AND NUTRITION

Carbohydrates, produced by plants, constitute the main source of food for the remainder of the living world, and as such form the basis of one of the world's largest industries. Production, distribution and nutritional aspects of foodstuffs, particularly the carbohydrate components, are therefore matters of the first importance. Food safety being of primary concern, the acceptance of products from genetically-modified crops is not universal as yet; nevertheless there is an urgent need to apply appropriate technology where food supply is seriously at risk, and there are other advantages of an economic or ecological nature in adopting the genetic route.

Bulk Foods and Vitamins

Carbohydrates comprise above 90% of the solid content of plants, which furnish starch and cellulose as well as a rich variety of more complex polysaccharides, and also sugars. Protein and lipid sources trace their origins from these substances. Microorganisms are vital in the food chains inasmuch as carbohydrates generally are utilized for growth, and the organisms enable ruminants to feed on grasses and forage from plant materials. The vast food industry can draw upon enormous natural reserves, mostly renewable through pHotosynthesis, and carbohydrate raw materials are used to generate a variety of products in increasing quantities. The importance of diet in promoting health is firmly accepted, total intake, the relative proportions of the bulk components, and the incorporation of adequate fibre and vitamins (among which is the sugar-derived antiscorbutic vitamin C) being of primary concern.

Food has to be palatable in every sense to each individual, and non-injurious to all, so the industry is of necessity under the watchful care of legislating bodies;

there is a problem here as habits and requirements vary greatly across the globe and so does the strictness with which food laws are applied. It is at this point that the cardinal role of chromatograpHic analysis in setting standards, and providing a sound basis for corrective action to be taken, may be perceived.

One of the most critical aspects of food management being preservation, monitoring and delay of onset of deterioration are alike considerations that concern the analytical chemist, the food scientist and the microbiologist. Because of the nutritional importance of carbohydrate, fermentation, spoilage and the deleterious activity of bacteria and fungi are particular dangers to be avoided and overcome. Pasteurization, hermetic packaging, refrigeration, addition of allowed preservatives, and scrupulous attention to handling and marketing procedures are among the standard procedures to be followed; the HACCP system recognizes that a mutidisciplinary approach to food safety is essential.

Heat treatment, the usual procedure, is to be minimized in such cases where the product is liable to develop undesirable taste, and apart from certain prejudices against the sterilization of foodstuffs by radiation treatment, there is always the question as to how the texture and integrity of polysaccharide-containing products are to be maintained under different conditions. Sugars and vitamins are particularly affected, and of all the many changes that occur during cooking the Amadori rearangement (Maillard, non-enzymatic browning reaction, described in 1912), involving sugars, amino acids and proteins, is perhaps the most significant.

It is appropriate here to introduce some of the terms associated with the nutritive properties of polysaccharides present in foodstuffs, and a quantitative comparison with the properties of other types of food constituent. The intrinsic nutritional value, digestibility and digestible energy, "food energy", metabolic energy yield, and safety in use are all expressions and factors that provide material for an ongoing debate. They fluctuate according to the food recipient, whether man, animal or poultry, cooking procedures, and the "food environment" which includes other commodities present apart from carbohydrate, and differ to a small extent from heats of combustion as determined by calorimetry.

The range of values for available nutritional energy varies from ca.l kJ g-1 (a "low-calorie" food) to ca.17 kJ g-1 for starches (which varies, portions of starch being "resistant" to digestibility- itself a term of wide interpretation), a little less (ca.16) for cross-linked or other modified starches, a maximum of 12 kJ g-1 for inulin (though the digestibility is low), ca.5 kJ g-1 for non-starch plant polysaccharides (NSP) and as an upper limit for dietary fibre generally; gum arabic (ca.70% or more digestible) and other plant gums vary from 5-11 kJ g-1, and alginates have "no nutritional value".

By way of comparison, lipids are about twice as "energetic" as starch; they may contribute nearly one-half of the total caloric intake of persons in developed countries, whereas 30% might be regarded as preferable. Proteins, apart from their great importance as nutritive substances, afford ca.17 kJ g-1 of energy, the same as "carbohydrate". Adult human requirements are of the order of 10MJ per day in total, three times as much as for a 6 month old child. Pet foods have not been neglected in this type of study.

Cellulose as Nutrient

Not only is cellulose the most abundant organic substance, but it is the major carbohydrate nutrient of herbivores and insects, which comprise the bulk of the animal kingdom. For humans, the cellulose component of grain, vegetables and fruit, or present in the diet as a supplement, is a valuable source of insoluble, dietary fibre.

Modification of cellulose by methyl ether formation is an effective way of increasing the solubility to the extent that incorporation into processed foods to promote thickening and gel formation, and to modify sensory perception of flavour, is both feasible and profitable. The variety of modifications and properties, and the types of food product to which refined and derivatized cellulose are commonly added, form the subject of reviews.

Starch

Laid down in the form of characteristic granules, starch is the chief reserve nutrient in plants, and because of its palatability and digestibility after cooking it forms the largest bulk nutrient and source of energy in human diets. Some starch, which resists amylolysis in the upper digestive tract, is a minor part of the ingested whole, but nevertheless an important dietary fibre source because of the large amounts of starch eaten. The uptake of starch as a food component is preceded by gelatinisation and pasting processes in the cooking of primary sources or in food technology, pHenomena that are of paramount importance, and which, among others relating to properties and preservation, continue to be studied in depth.

Equally significant are the many food products obtained from starches used as raw material, such as dextrins and syrups from partial hydrolysates, cross-linked and other esterified derivatives, hydroxypropylated starches, and even partially oxidized modifications. The object is value enhancement through converting the starch to more soluble, stable, or more easily manipulated products in conformity with modern demands of the food industry and of the consumer public. One such requirement is the provision of foods with lowered lipid (fat) content; to a marked degree modified starch hydrolysates can serve this purpose and are widely included in commercial fat replacers. A comprehensive review is available to enlarge on many technological aspects of the starch industry.

Sucrose (Sugar)

Sucrose is the other carbohydrate consumed on a gigantic scale worldwide. Known universally as sugar, sucrose is highly nutritive, has a glycaemic index of 65, and is the predominant sweetening agent. An early product of pHotosynthesis, and stored in easily accessible tissues of many plant taxa, sucrose is extracted commercially from crushed sugar cane in subtropical to tropical regions and from sliced beet in temperate zones, the two sources in the approximate ratio.

Calcium hydroxide is added to the juice from cane, the solid cake is removed by filtration, and the crystalline disaccharide is recovered in remarkably pure form by repeatedly concentrating the solution under reduced pressure and removing successive crops. Molasses is the syrupy residue, and the sucrose product is refined as required by the type of food use to which it is to be put. Beet sugar is obtained similarly, together with small amounts of the very much less-sweet oligosaccharides raffinose and stachyose. Lactose and palatinose, an isomerisation product from sucrose, are other disaccharides accessible on a large scale.

Sucrose is an amazingly versatile starting point for conversion to synthetic products; the term sucrochemistry, coined half a century ago, reflects the special importance attached to the technology involved. Among many valuable derivatives of sucrose is the water-soluble galactosyl analogue with three of its hydroxyl groups substituted by chlorine known as sucralose (4,l', 6'-trichloro-4, l', 6'-trideoxy-galacto- sucrose). Being more stable in acid and having a sweetness over 600 times that of sucrose, there are many uses for this substance in the countries granting approval for food use. Aspects of the perception of sweetness and of bitterness, and the many factors to be considered in replacing sucrose as a food ingredient. Sucrose esterified (DS >5) with fatty acids is not metabolized or absorbed and yet possesses the same taste and cooking properties as normal fats (glycerides), so that its development as a low-energy substitute has been profitable.

Selective formation of derivatives is not an easy undertaking, owing partly to the labile nature of the glycosidic bond. Sucrose monostearate is an acceptable emulsifier in food products. Amounts of sucrose even larger than those consumed in foodstuffs by the populations of developed countries, form the basis of the vast fermentation industry, which includes the production of antibiotics, baked goods, and animal pet foods. Of the monosaccharide group D-glucose is produced in amount equal to about 4% of that of sucrose, approximately one-eighth of which is converted by reduction into sorbitol and increasing quantities are isomerized to D-fructose.

Vitamins

Vitamin C, ascorbic acid, is the predominant vitamin of carbohydrate (L-sorbose) origin, and is a member of the water soluble group. The unusual unsaturated 1, 4-lactone structure permits the formation of, for example, a monosodium salt derived from the enediol system, which in turn underlies the reducing capability of the molecule. Great industrial importance is attached to this substance, which as an antiscorbutic agent is indispensable and as a food additive is a widely used antioxidant with ancillary antimicrobial properties.

Vitamin C is essential for the bodily synthesis of connective tissues, and is widely recognised as a preventative agent in the treatment of the common cold. Riboflavin (vitamin B2 or G; D-ribitylflavin) is a nutritionally important enzyme cofactor which is widespread in plants and animals and concentrated in yeast. Vitamin L2 is a thio derivative of adenine, and an essential moiety of cyanocobalamine (vitamin B12) and some of its congeners is the -D-ribofuranosyl group. This critically important vitamin occurs exclusively in foods of animal origin, and may be manufactured using certain bacteria only. There has been a surge of interest in the health benefit of what are termed glyconutrients — the group of monosaccharides comprising the glycoconjugates. To these are added in nutritional supplement formulations a wide range of natural polysaccharides that contain the sugars. About three-quarters of the population of developed countries take multivitamin and mineral supplements, with some 10% reporting adverse effects, though the general view is that supplements are necessary.

Glucosamine and extracts of Echinacea, garlic, ginseng and ginger are among the most popular. Whether the value of glucosamine (and chondroitin sulpHate) lies in their ability to stave off arthritis or if they are rather to be seen as "nutraceuticals" is a matter that is under investigation.

FOOD HYDROCOLLOIDS

These polysaccharides show a range of pHysical properties that make them invaluable, as additives, in manufacturing most of the range of foodstuffs offered by the industry. All are available from plant or microbial sources, or can be produced in cell culture, and are usually isolated or extracted in a suitable form without any great difficulty. Derivatization is commonly carried out in order further to improve properties for special needs; irradiation likewise produces various effects.

The carbohydrates classified as hydrocolloids are sometimes deemed not to include starch, but because of the wide range of starch derivatives having properties in common with the rest, and because of the extremely large quantities involved, starches are grouped with cellulosics, other glucans, mannans, inulins, pectins, bacterial gums, plant gums, and the pHycocolloids, The food usage of starch probably exceeds that of the others by a factor of 5 or more in Europe; though on a weight for weight basis the non-starch polysaccharides are about 8 times more valuable.

These hydropHilic polymers share such common characteristics as the ability to produce viscous dispersions under a variety of conditions, and in many instances to form gels even at concentrations as low as 1%. A large molecular size is essential, structure and conformation determining the properties of individual polysaccharides. Linear molecules occupy larger volumes than branched molecules of the same molecular weight, and the flexibility of chains permits the adoption of expandable, random coil conformations in solution, with corresponding rheological effects.

When segments of these molecular chains form intermolecular bonds, a system is created that is able to entrain large volumes of water, and gels result. Even if there are no ionizable groups or hydropHobic regions an envelope of hydroxyls is always present to interact strongly with water in the medium by hydrogen bonding, as well as with the wide range of other substances (this includes self-association) necessarily present in the same solution when operating under practical conditions. A general feature is the low concentration (often <0.5%) at which hydrocolloids manifest their action as permitted additives, and the strong influence they exert upon the food systems concerned.

Extensive researches have been made into the correlation of molecular structure with gel-forming and other important rheological properties, and in particular the synergy between different polysaccharides in various mixtures; interactions between xanthan and galactomannans in solution represent a well-studied case. Interactions with protein are of prime importance also in emulsification and the stabilisation of oil in water emulsions, as well as in film formation.

The number and variety of applications is both large and increasing. A few of the uses to which food hydrocolloids are put, it is advisable to consult the literature and proceedings of the many conferences on this subject, and the journal Food Hydrocolloids, to obtain a fuller picture. The question of food preservation and the influence of hydrocolloids on the quality of various foodstuffs are subjects of the greatest importance to manufacturers and consumers alike. The identification of gum additives in processed foods is conveniently achieved using antibodies. A new concept is the possibilty that the cell-signalling potential of hydrocolloids could be of technical interest.

Dietary Fibre

The concept of dietary fibre has many facets and is closely linked with considerations of energy provision, so that a complete description of the various aspects of nutrition will necessarily include both of these factors. Carbohydrates, as has been pointed out, are the main sources of nutrient energy in the animal world, and within the category starches constitute the major single component. The heat of combustion of starch (17.68 kJ g-1) is a useful reference value, converting to an equivalent energy of 15.8 for glucose, the form in which starch is used up in metabolism.

A fraction only (though a large one) of starch is fully digested and absorbed in the small intestine, which portion is referred to as available starch. Unavailable starch and non-starch polysaccharides are to an extent fermentable in the lower intestine, by bacterial action, yielding some energy, the precise amount of which is difficult to determine in practice. Factors that affect the digestible energy of unavailable complex carbohydrates in foods include the levels of such components in mixed diets and the form in which they are ingested, but the general conclusion reached is that the average value approximates to 0.5 kJ g-1; for some types of dietary fibre, a figure of 3 may apply.

Non-starch polysaccharides, starch derivatives and resistant starch constitute the bulk of dietary fibre, the remainder being mostly lignin and indigestible protein. For plant-based foods, this is essentially the set of components of the cell-wall, including pectic substances, and other acidic polysaccharides such as exudates. Included also are hydrocolloids of algal or bacterial origin. Details as to the types of compound involved have already been presented in discussing rhamnogalacturonans, xyloglucans, and glucuronoarabinoxylans in addition to the major structural entity cellulose; at least seven different monosaccharide units are involved, and the arabinogalactan-proteins introduce another chemical variation. Seeds, storage organs, fruit and vegetables are all sources of the cell wall polysaccharides consumed by humans.

Definition of the term dietary fibre itself is far from straightforward, but in order to perform chemical analysis for reporting and comparison purposes it is required that the components should be specified and that the methodology should be adequate for the task. This has its limitations when the need exists for an assessment of the contribution of the fibre to the various, relevant pHysiological processes; here a functional approach might be preferable, such as the measurement of glycaemic index, water-holding capacity or effect on viscosity.

In practice a number of accepted methods have been devised for the determination of dietary fibre and the respective proportions of insoluble fibre and soluble (the lesser component, but considerably more prominent, approaching one-third, in vegetables than in cereals). Common to most of the methods, which approximate to the natural digestive processes, is the initial removal from the food sample of both starch and protein by enzymatic treatment or by suitable solvent extraction, followed by gravimetric and compositional analysis of the residue..

Regarding the importance of dietary fibre (in its broad definition) as an integral part of the diet, it should be noted that the soluble portions, consisting of polysaccharides leached from ingested cell-wall components, thickening agents and stabilizers, exert their effects, such as viscosity enhancement, near the commencement of the upper digestive tract. One important consequence of this is the delaying and lessening of any increase of blood glucose concentration during the first hour after eating; insoluble fibre has little or no effect in this respect. The therapeutic use of fibre has long been an important component of the management of diabetes, the aim being to maintain blood glucose levels at about 5-9mmol L-l.

This treatment is usually successful in the early years after diagnosis of the non-insulin dependent (type II) condition. Guaran added to the diet is comparable in its effect to polysaccharides from oats and some leguminous seeds. Such soluble polysaccharides, as well as pectins, are similarly able to cause lowering of both total and low-density lipoprotein cholesterol in serum (particularly LDL cholesterol); there is a link here with the lowered incidence of coronary heart disease, though whether the diet as a whole or the non-starch polysaccharide fraction provides the protection is debatable. Mineral levels may also be affected by dietary fibre through the cation-binding capacity of, inter alia, the carboxylic acid groups present in galacturonans.

The pHysiological effects of insoluble fibre include protection against the large bowel diseases diverticulosis (oatmeal and wheat bran being favoured) and colorectal cancer (wheat bran is indicated from studies using chemical carcinogens on animals, and vegetable fibre from epidemiological evidence). Both xyloglucan and acidic rhamnogalacturonan have been demonstrated as showing inhibition of mutagenic activity in a microbiologial testing screen, and this may explain some of the cancer-preventing effects of high-fibre diets. The water-holding capacity of the fibre in the hindgut and its functioning as a bulking agent on the one hand, and the extent to which the fibre acts as a carbon source for increased microbial growth in the colon on the other, are considered to be the factors involved in the protective action of the dietary fibre.

In addition to these specific instances, dietary fibre is believed to have a protective role against many other, multifactorial diseases and undesirable conditions including gastric and duodenal ulceration, gallstones and renal calculi, hiatus hernia, haemorrhoids, inflammatory bowel diseases, deep vein thrombosis, constipation and obesity. The presence of certain microorganisms in the colon is regarded as beneficial to health (as "probiotics"), lactobacilli and bifidobacteria for example, and their growth is promoted by the availability of dietary items that pass through the small intestine ("prebiotics"; contrast the use of this word to describe simple compounds associated with primitive cosmic conditions!).

Included among these are lactulose, galactooligosaccharides, inulins, and oat products; other saccharides having effects on bowel action, not always desirable, are lactitol and the sucrose-related raffinose and stachyose. Certain individuals suffer from lactose intolerance, consequent upon a deficiency of intestinal lactase and the passage of the disaccharide into the colon, where bacterial fermentation takes place and abdominal discomfort is experienced.

Sweeteners

The superior qualities and abundance of sucrose place this disaccharide in a class of its own as a sweetener and as an auxiliary food ingredient, providing bulkiness and texture support in bakery products, acting as a preservative, and at a concentration usually of about 10% forming the most acceptable, nutritive taste provider in beverages of all kinds. The energy released on metabolism is comparable with that of glucose, but the sweetness is higher on account of the fructose moiety present. Apart from the application of sucrose in purified, crystalline form, the commercially available syrups from hydrolysed starch, molasses and maple exudate are the principal forms in which sucrose, glucose and fructose are employed.

There are, however, disadvantages to the use of sucrose as a sweetener. In common with other sugars and fermentable carbohydrates generally, sucrose provides a growth medium for microorganisms in the mouth, resulting in tooth decay. Such are the increasing demands of society in many parts of the globe, that alternative sweeteners are being developed and investigated on a large scale in order to provide substances that are in fact non-nutritive, as well as being non-cariogenic. Inasmuch as products are available that are sweeter than sucrose by very large factors, and need be applied in correspondingly smaller amounts, these considerations are easily satisfied, but the need remains to compensate for the loss of the auxiliary properties of sucrose if substitutes are used.

Sweeteners other than Sucrose

Alternative nutritive sweeteners include the whole range of polyols formed on reducing-end conversion to CH2OH by catalytic hydrogenation of glucose, erythrose, xylose, maltose, lactose, or maltooligosaccharides. These products are generally less sweet than sucrose and about two thirds as "caloric", to use a commercial term, though erythritol is only one tenth as energy yielding. The synthetic polysaccharide termed polydextrose is a bulking agent with sweetener properties.

Among monosaccharide alternatives D-fructose, 50% sweeter than sucrose, has the advantage of not causing serum glucose levels to rise after ingestion, while the fructooligosaccharide group (from fructosan hydrolysis) is both one half as sweet as sucrose and one half as caloric. Fructose is the principal (38%) component of honey, which has been been used since prehistoric times and still enjoys great popularity. Glucose (31%) and water (17%) are the other main constituents, reflecting both the "inversion" of sucrose during passage through the female honey bee and the monosaccharides present in the nectars ingested. A rich variety of di- and oligosaccharides (about 10% of the honey by weight, on a dry basis) have been revealed by methods of chromatograpHic analysis. Honey dews are insect (apHid) products, found on leaves and other parts of many trees, which may be transformed by bees when nectar is not readily available into honey-dew honey, in which the proportion of maltose is high (9%) and the characteristic trisaccharides melezitose and erlose are found. Phloem sap may be yet another nutrient source for bees in the production of honey.

High-fructose corn syrup, obtained by enzymatic (isomerase) conversion of syrups containing glucose (90-95% on a dry basis) to an approximately 50% mixture, is a popular form in which fructose can be administered. It is possible to raise the fructose content substantially. Lately, D-tagatose, the lyxo counterpart of fructose, has been accorded GRAS status as a sweetener, with 30% of the nutritive value and 92% of the sweetness of sucrose. It is potentially abundant as an alkali-isomerized derivative of galactose from whey. The term "non-nutritive" as applied to sweeteners is no reflection on the "caloric" value of the sweetener per se, but is used in the sense that traces only are normally administered.

The six principal non-nutritive sweeteners used commercially are aspartame (a peptide derivative, based on pHenylalanine), saccharin (sulpHimide), acesulfame-K (oxathiazine derivative, K salt), and cyclamate (cyclohexa-nesulpHamic acid, Na salt), which are not carbohydrates, and the little-used stevioside (diterpene glycoside) and glycyrrhizin (triterpenoid glycoside of a B-2-linked disaccharide comprising D-glucuronosyl units, neutralized) from leaves and roots respectively. L-Sugars are potential bulking agents of the non-nutritive class.

Adding to the wide diversity of organic substances having sweetening properties, some of which are being discovered by screening of natural resources while synthesis is continually generating countless others, are the astonishing range of deoxyhalo derivatives of sucrose having relative sweetness values of between 5 and 7500, methylthio derivatives, other modified aspartames (up to 50000 times sucrose) incuding alitame (2000), the chalcone glycoside neohesperidin dihydrochalcone (300), and sweet-tasting proteins thaumatin ((2000) and monellin (2500).

At the top of the scale at present is the substance sucrononic acid (200000), developed on the basis of molecular modelling of the structural elements understood to be necessary for the perception of sweetness. This success is in itself a proof of the efficacy of attempts to correlate structure with a sweet response. Some unmodified amino acids, D- or L-, are sweet; a synthetic compound, 1000 times as sweet as sucrose, is an interesting example of a substance composed of carbohydrate and amino acid moieties, the sugar having a bromine substituent to function as an auxiliary hydropHobic group necessary to intensify the sweetness. It should of course be noted that substances quite unrelated to any of those described, even certain lead salts, are sweet to the taste.

Sweet Taste

What is certain is that the apprehension of taste can at best be analysed aspect by aspect, coordination of all factors to understand the sense in its entirety not yet being a possibility. Concentration effects, synergism and antagonism involving other food and salivary constituents, mouthfeel sensations, after taste, genetic considerations and even the mood of the taster all combine to the highly complex problem of defining the perception and performance of a sweetener.

The study of structure/activity relationships using sugars and other sweet-bitter stimulants as models dates back to the work of G. Cohn published in 1914, organic functional groups, often occurring in pairs, being classified according to their respective abilities to induce a sweet or a bitter taste. Stereochemical differences were in due course seen to produce dramatic changes in both of these responses. It finally became clear that the initiation of a sensation of taste must involve the interaction of a stimulant molecule located on the taste-cell plasma-membrane, the action occurring in the taste buds of the tongue; from this point impulses reach the brain and are "interpreted".

The role of hydrogen bonding in sweetener-receptor interaction was recognised by Shallenberger, who concluded that sugar sweetness varied inversely with the extent to which intramolecular hydrogen bonding could occur, though intermolecular bonding might also be relevant. The concept that in a sweet molecule a proton donor A(H) should be sited at an optimum distance of 300pm from a proton acceptor B, with complementary atomic groups on the receptor to which hydrogen bonding could occur, has proved to be valid for many other classes of compound as well, including saccharin, cyclamic acid, L-alainine and nitroanilines.

The next pHase of understanding implicated a third, hydropHobic binding site for a potent sweet-taste response, designated, delta, or X for sugars, dipeptides and. nitroanilines respectively. The location of the site with respect to AB varies, but in each instance forms an unsymmetrical (scalene) triangle; in the case of sugars A is about 314pm, B 525pm and angle BA 103°. A detailed analysis of the structures and conformations of a host of sugars, methyl ethers, anhydro sugars, oximes, glycosides, (l,2)-linked disaccharides and other compounds in terms of the AH, B, concept and sweetness appears in the review by C-K. Lee.

The unique situation regarding the intensely sweet response shown by the deoxyhalosucroses, whereas all such derivatives of other sugars are extremely bitter, may be the consequence of the sweetness-generating system spanning both glucose and fructose units and being able to lock on to the complementary site. This group of powerful sweeteners illustrates the synergistic action of two hydropHobic blocking sites and the dramatic influence of the size of the halogen substituents.

Alternative Taste Sensations

Bitter taste is related to lipopHilicity, as shown by the properties of dideoxysugars, sugar acetates, and glycosides with long carbon chains as aglycones. As regards carbohydrates that have any taste at all, some may show both of the fundamental tastes by interacting with two different receptor sites within a receptor location. Sweetness can be inhibited by derivatization of synthetic carbohydrates, and by the food-allowed salt sodium 2-(4-methoxypHenoxy)-propanoate which is marketed as the racemate (lactisole), though only the S-(-)-enantiomer is active.

Because the inhibition in this case affects the sweet taste of both sucrose and aspartame it might be argued that the existence of a sweetener receptor is a reality, but if so its nature is far from clear. An extensive model relating structure and sweetness has nevertheless assumed the possible implication of no fewer than eight recognition sites.

It has been postulated that the binding of sugars to receptors triggers a secondary messenger cascade involving inter alia an important GTP-binding protein, cyclic AMP and a dependent protein kinase, a resulting influx of Ca^{2+} ions, cell-depolarization, and neurotransmitter release. Neural signals encoding sweet taste information are then carried to the central nervous system. Likewise saccharin has been shown to enhance the production of inositol 1,4,5-trispHospHate, With a release of Ca^{2+} from storage depots, and similar consequences.

INDUSTRIAL

Usage of Polymeric Carbohydrates, Derived Sugars and Natural Oligosaccharides

Enough has been written about the variety and quantity of terrestrial carbohydrate to show that the present and potential availability of raw material of this type for industrial use need not cause concern: irresponsible and wasteful practices could, however, alter the position in due course.

Cellulose, derived mainly from timber by delignification and the removal of solubles during pulping processes, is converted largely into paper and textile products, and after purification forms the substrate for ether and ester products which have a plethora of uses. Enzyme treatment is an effective preliminary to derivatization.

Carboxymethyl cellulose is a dominant derivative, used as a laundry detergent and more particularly a food additive, though the enormous part played by cellulose nitrates in the explosives and plastics industries, and by acetates in fibre formation, should not be discounted. The potential of chitin is not yet fully realised, in spite of considerable attention being paid to its medicinal properties, but processed starch and other soluble polysaccharides of plant origin, gums, and fungal and bacterial products find wide application as industrial thickeners and sizes in oilfield operations, adhesives, detergents, explosives, surface coatings, paints, and many divisions of the foodstuff industry, even the manufacture of edible plastics. The section on hydrocolloids explains how their pHysical and functional properties are related. Starch and sugars are furthermore basic requirements in the fermentation industries.

Sucrose occupies a uniquely important position as a renewable, bulk commodity (more than 108 tonnes used annually), available as concentrated syrups or in a highly refined form. The low cost of production and high yield in terms of land use combine with versatility of properties as a nutrient, sweetening agent, and growth medium for yeasts and microorganisms generally, to give this disaccharide its special status.

However, despite the almost limitless possibilities for functional group modification, not many derivatives, apart from sweeteners and detergents, and some food additives, have entered the market in processed form. An interesting enzymatic conversion of sucrose to to the less sweet and less digestible cellobiose has been effected via pHospHorylation to glucose 1-pHospHate, isomerisation of the concomitant fructose, and coupling.

Much the same difficulty, namely the limiting and control of derivatization of the hydroxyl groups, applies to lactose and the large quantities of monosaccharides, and from them polyols, that are available on hydrolysis of higher saccharides. This problem will be overcome to the extent that improved synthetic methodologies are applied in the future. Controlled pyrolysis of carbohydrate raw materials and adaptations of the purely thermal approach provide other rich sources of commercial organic chemicals.

Detergents

Detergents are substances that perform cleansing functions, often as mixtures of organic surface-active agents (surfactants) with other substances added to improve efficiency and acceptability. Surfactants are composed of molecules containing moieties of opposite solubilities, typically a hydropHobic, lipopHilic hydrocarbon of considerable chain length (up to ca.20 methylene groups), and a hydropHilic, often ionic, water-soluble grouping.

Adsorption and molecular orientation take place at interfaces of the liquid system in which the surfactant is placed, and if the concentration is high in the bulk of the solution, surfactants form aggregates, calxled micelles.

Surface tension lowering at interfaces results in highly desirable combinations of properties including detergency, wetting, foaming, emulsification, film-formation and solubilization. In most instances the medium containing opposite types of pHase is predominantly aqueous, and certainly until a century ago the only synthetic surfactant was soap, the alkali metal salt of long chain carboxylic (fatty) acids.

Shortcomings such as hard-water insolubility resulted in the development instead of sulpHates and sulpHonates in which many varieties of hydropHobic groups were integral; the raw materials for these detergents included products derived from petroleum to add to those from animal and plant fats and oils. The hydropHobic moiety might contain any of a great variety of atoms other than carbon, unsaturation, or functional groupings, one of the general features being the extent of polydispersity tolerated by the surfactant concerned.

Carbohydrate-containing surfactants, of the non-ionic class, do not, as yet, occupy a particularly large share of the vast range of modern products devised to satisfy the wide range of commercial requirements, but for example there is a large class of carboxylic ester surfactants which are derivatives of the polyhydroxylic 1, 4-sorbitan and the dianhydride of sorbitol, variously esterified with hydropHobic moieties. Perceived as "natural" these surfacants have a certain advantage.

Appearing under numerous trade names (Armotans, Arlacels, Emsorbs, Spans and others) these fatty esters are soluble in mineral and vegetable oils, not in water, and as lipopHilic emulsifiers find application in textile manufacture and processing, and in cosmetics and food preparations.

More-hydropHilic derivatives of the anhydrosorbitols are formed by reaction with an excess of ethylene oxide to yield complex mixtures containing chains of ethoxy groups, monoesterification of which with stearic (or a related) acid gives another set of products, the polysorbates. These include ICI's Tweens with emulsifying, wetting and solubilizing properties. Monosubstituted sucrose esters of fatty acids have been mentioned as another example of a sugar-derived product having emulsifying properties, some being for specialist, biomedical purposes.

Reaction of glucose with methylamine followed by double bond reduction produces a substituted glucamine, which when N-acylated with fatty acid methyl esters yields fatty acyl glucamides, sold in the form of dishwashing liquids. As in the case of sorbitan derivatives, products modified to include polyoxyethylene groups have proved to have excellent surfactant characteristics. Substituted galactosylamines, analogous in certain respects, have been investigated.

Aminolactitol, N-alkylated with long alipHatic chains, has the property of associating with dibasic acids to form self-assembling catanionic surfactants, of considerable interest as anti-HIV agents.

N-Substituted derivatives of glucosamine with a steroidal hydropHobic moiety, resembling the steroid saponin digitonin and similar in action to detergents of the alkyl glycoside type, have recently been synthesized for the solubilization of membrane proteins. n-Alkyl glycosides are among the carbohydrates known to be liquid crystalline compounds, and have both ampHipHilic and mesogenic properties; galacturonic acid derivatives of several kinds have similar characteristics, and sucrose ethers are now available as ampHipHilic substances.

BolaampHipHiles have a hydropHobic core (such as a long alkyl chain) with hydropHilic moieties bonded at both ends. These compounds have the property of self-organization in water to form stable monolayer lipid membranes, useful as models, and having biomedical applications. The unusual properties of surfactants of this type have stimulated the synthesis of a variety of structural classes based on carbohydrates, including gemini surfactants which have the property of forming thread-like, as opposed to spHerical, micelles. These products are ampHipHiles possessing in sequence a long hydrocarbon chain, an ionic (hydropHilic) grouping, a spacer, then a second hydropHilic group followed by one that is hydropHobic. Even polysaccharides may serve as hydropHilic head groups, xylans and cellulose derivatives for example; in the instances cited both the xylan and the cellulose were rendered ionic and therefore more soluble by derivatisation, changing thereby the usual, nonionic character of carbohydrate-based surfactants to ionic in these cases. From dextran, ampHipHilic derivatives have been synthesized.

Adhesives

Adhesives cause the attachment of solids to each other by acting on their surfaces through various mechanisms, such as diffusion, electrostatic attraction, or the operation of intermolecular forces. Carbohydrates and proteins are the main natural sources, though today most adhesives are synthetic organic products of many different types. The water-soluble carbohydrate gums, have been used since antiquity, the effectiveness of these substances being attributable to their high molecular weights, hydrogen-bonding potential, and in varying degree to ionic properties. Apart from the gums, modified starches have long been employed as industrial adhesives, notably acid-treated dry or wet starch (dextrins white and yellow, pyrodextrins, British gums), oxidized starch, and various esters. Among the cellulosics, acetone-soluble nitrate esters were the first synthetic adhesives developed (in 1869), since when a wide range of other esters and ethers have assumed prominence. In total, however, carbohydrate-based adhesives enjoy only a minor share of the market.

BIOMASS AND ENERGY

Carbohydrates being the most abundant of organic chemical types, accounting for well over one-half of the total carbon present in combined form on Earth, the bulk of renewable matter available for conversion to energy, use in building construction, food provision and transformation into value-added products, comes from this source. Preserving an acceptable balance between natural forces on the one hand, and the depredations caused by the world's population and its industrial activity on the other, is undoubtedly a matter of the gravest concern.

Industrial Applications Agricultural and other Natural Products

Although cellulose, the hemicelluloses, starches and sucrose are much utilized in providing food, there is an ample surplus of polysaccharides and sucrose for industrial use, though clearly serious distribution problems exist just as they do for foodstuffs and much else. To these major commodities must be added the less accessible though plentiful carbohydrates; chitin is a prime example, together with untapped resources of the sea in the form of marine plants, animals and plankton, which are less abundant.

Chemurgy is a term used to describe that branch of applied chemistry which utilizes plant or microbial material to form diverse products of enhanced value or to generate energy. The review cited indicates the means adopted in the United States to achieve these ends from such renewable sources as industrial crops (of which trees form the bulk), surplus food crops, and wastes obtained from these raw materials.

The particular contribution of biomass to the provision of energy has been comprehensively reviewed, the main focus being upon materials containing fixed carbon (organic compounds) that are renewed frequently enough to make them continuously available in large quantities to maintain and augment energy supplies.

This clearly implicates carbohydrates, and as clearly excludes fossil reserves that take millions of years to be replenished. Estimates of the global distribution of biomass formed pHotosynthetically show that whereas the total produced terrestrially is double that from the sea, the available amount is perhaps 200 times as great on account of a much lower turnover rate. Tropical forests and to a lesser extent temperate deciduous and evergreen trees constitute practically 90% to the total of standing biomass, grasses, such unwanted material as the insidious water hyacinth, and cultivated crops, providing the rest. The dismaying destruction of rain forests is accordingly a serious drain on biomass productivity, and although forests are presently only a minor contributor to energy provision, they are nevertheless essential for the preservation of life forms, and will assume growing importance as fossil sources are depleted.

The chemical composition of biomass varies, of course, as the individual components differ from that of the predominant cellulose, sludge and grasses contributing up to about 5% of N while the C content remains near the value for cellulose. From this reserve are obtained energy, directly by combustion or fermentation, or via fuel intermediates, and, using industrial procedures, chemicals of many types. Pyrolysis, hydrogenation, fermentations, hydrolysis, and oxidation processes are the most common, wood remaining the raw material of choice (hence the term applied to the products, siivichemicals; cf. sucrochemistry);

Alkaline degradation is a serious concern in cellulose technology. Pyrolysis of wood at ca. 500°C yields charcoal and gases (CO_2, CO, CH_4, and N_2 principally), tar, methanol, ethyl acetate and other organic solvents, and residual pitch, but after hydrolytic treatment wood affords glucose in sufficient yield (about 50% of theory) to provide a rich source of fermentation products. In a similar fashion such agricultural waste as corn cobs has long been submitted to chemical treatment, generating furfural (2-furaldehyde). Lichtenthaler has recently surveyed the composition of carbohydrate sources of biomass and this potential.

Pyrolysis Experiments

Fundamental studies of pyrolysis and other degradative processes have been made on a range of important biomass components, notably cellulose, starch, xylans, inulin and sugars, in order to gain the understanding of the mechanisms involved necessary to optimise and develop the methodologies practised. This has required the acquisition of purified and in some cases isotopically labelled carbohydrates, and sopHisticated use of chromatograpHic and spectrometric techniques. Pyrolysis of cellulose results in transglycosylation, dehydration, and subsequent decomposition and charring reactions, both homo- and heterolytic; major product differences result from variation in temperature and the presence or absence of acid, or base, or metal ions.

An early intermediate in the subsequent formation of the major low-molecular weight compounds glycolaldehyde, hydroxypropanone, and formic and acetic acids from glucans is considered to be 1, 6-anhydro-β-D-glucose (levoglucosan), the formation of which via an achiral C-l carbocation has been postulated on the basis of model studies. The presence of inorganic contaminants favours the production of these smaller molecules rather than anhydrosugar.

Decomposition of carbonaceous chars obtained from cellulose under relatively mild conditions has led to the identification of over 150 substances, the large majority aromatic or heterocyclic derivatives. Thermal oligomerisation and polymerization in the presence of an acid catalyst is a characteristic of methyl glycosides or sugars, and one such product, a mixed-linkage xylan, has been submitted to vacuum pyrolysis as a model for hemicellulosic polymers of this class.

Char yield was much greater than that for cellulose, and in addition there were produced 1, 4-anhydro-D-xylopyranose (contrast levoglucosan) and l, 5-anhydro-4-deoxypent-l-en-3-ulose (a suggested marker for xylans in biomass). Xylose itself, produced by hydrolysis, is a precursor in the industrial manufacture of furfural on an immense scale. The mechanism of formation of the corresponding homologue from fructose and sucrose (involving a fructofuranosyl cation), and numerous decomposition products, is considered in a seminal paper. Among the other key products of thermal decomposition of cellulose is levoglucosenone, like many other similarly-derived compounds a valuable starting material for chiral syntheses.

Novel Polymers

The abundance and stereochemical diversity of carbohydrates are leading to a corresponding development of new products, not only to the derivatives referred to above but also to degradation products that can be converted in turn to new types of polymer. A prime example is the synthesis of hydropHilic and biodegradable polyamides from monomers bearing amino and carboxyl functions, numbers of which have medical applications.

Given the correct additional substituents, such starting materials have been converted to film-forming, regular polymers that are both hydropHilic and chloroform-soluble, and conform to the requirements of biocompatibility. At the same time derivatives of cellulose, starch and chitin continue to be incorporated in blended polymer products that are thermoplastic and adaptable to special uses, as references to the patent literature will show.

THE FERMENTATION INDUSTRY

CELL CULTURE

Fermentation has traditionally involved the anaerobic evolution of carbon dioxide from fruits and malted grain by the action of yeasts and some bacteria, yielding ethanol in particular, but the term now includes aerobic processes with an extraordinary range of important and valuable products.

Those marketed are limited to only a few hundred in number, but the potential for diversification now that the production of genetically-modified microorganisms is growing dramatically, is great. Industrial ethanol, alcoholic beverages, yoghurt, soy sauce, various organic acids [acetic, lactic, citric, glutamic, gluconic; lysine and pHenylalanine], alcohols and ketones, trehaloses, cyclodextrins, enzymes [amylases, glucose isomerase, and lipases], vitamin B12, hormones and a wide range of antibiotics typify the scope of the fermentation industry today.

The isolation of products from the broth, which includes carbohydrate as nutrient in the form of molasses, grain, starches and sugars, cellulose hydrolysates being an important source of these, is crucial to successful production, as concentrations of the required compounds range from 5-15% or to as low as 0.1% Solvent extraction is common, whereas chromatograpHy is as usual reserved for analysis and the final purification of rare and valuable products. The degradation of cellulose and hemicelluloses is a natural process of consequence, and there have been many commercial applications of the use of fungi to produce ethanol and other organic solvents.

Although cell culture is no new process and has long been the basis of many studies of plant metabolism, a preparative tool, and an indispensable adjunct to cancer research, a relatively modern development is the large scale culture of animal, insect and plant cells for the purpose of generating proteins of value and viral vaccines. As techniques are adapted to producing these types of cell, which are more fragile (not having cell walls) than those from microorganisms and require special conditions for their their growth (which is also comparatively slow), there are many possibilities for future diversification. Monoclonal antibodies, erythropoeitin, and various "factors" against coronary heart attacks, haemopHilia, septic shock and transplant rejection are already available. Carbohydrates are essential nutrients for cell growth, and in the case of attachment-dependent cells, supports are needed, dextrans being serviceable in this respect.

CHAPTER 4

Carbohydrate

Carbohydrates (from hydrates of carbon) or saccharides are the most abundant of the four major classes of biomolecules, which also include proteins, lipids and nucleic acids. They fill numerous roles in living things, such as the storage and transport of energy (starch, glycogen) and structural components (cellulose in plants, chitin in animals). Additionally, carbohydrates and their derivatives play major roles in the working process of the immune system, fertilization, pathogenesis, blood clotting, and development.

Chemically, carbohydrates are simple organic compounds that are aldehydes or ketones with many hydroxyl groups added, usually one on each carbon atom that is not part of the aldehyde or ketone functional group. The basic carbohydrate units are called monosaccharides, such as glucose, galactose, and fructose. The general stoichiometric formula of an unmodified monosaccharide is $(CH_2O)_n$, where n is any number of three or greater; however, many molecules with formulae that differ slightly from this are still called carbohydrates and other compounds that possess formulae that agree with this general rule may not be in fact carbohydrates (eg formaldehyde).

Despite the inexactness of the term, "carbohydrate" remains a useful descriptive name and with a little experience even a novice will soon become aware of what is, and is not, a carbohydrate. Monosaccharides can be linked together in almost limitless ways. Many carbohydrates contain one or more modified monosaccharide units that have had one or more groups replaced or removed. For example, deoxyribose, a component of DNA, is a modified version of ribose; chitin is composed of repeating units of N-acetylglucosamine, a nitrogen-containing form of glucose. The names of carbohydrates often end in the suffix -ose.

MONOSACCHARIDES

D-glucose is an aldohexose with the formula $(CH_2O)_6$. The red atoms highlight the aldehyde group, and the blue atoms highlight the asymmetric centre furthest from the aldehyde; because this -OH is on the right of the Fischer projection, this is a D sugar.

Monosaccharides are the simplest carbohydrates in that they cannot be hydrolyzed to smaller carbohydrates. The general chemical formula of an unmodified monosaccharide is $(CH_2O)_n$, where n is any number of three or greater.

CLASSIFICATION OF MONOSACCHARIDES

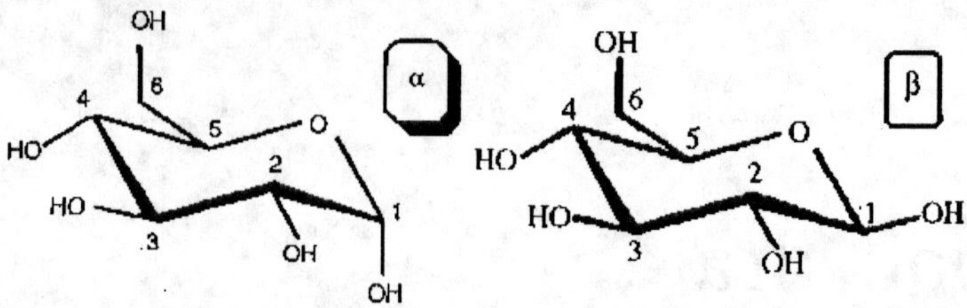

The α and β anomers of glucose. Note the position of the anomeric carbon (red or green) relative to the CH_2OH group bound to carbon 5: they are either on the opposite sides (α), or the same side (β).

Monosaccharides are classified according to three different characteristics: the placement of its carbonyl group, the number of carbon atoms it contains, and its chiral handedness. If the carbonyl group is an aldehyde, the monosaccharide is an aldose; if the carbonyl group is a ketone, the monosaccharide is a ketose. Monosaccharides with three carbon atoms are called trioses, those with four are called tetroses, five are called pentoses, six are hexoses, and so on. These two systems of classification are often combined. For example, glucose is an aldohexose (a six-carbon aldehyde), ribose is an aldopentose (a five-carbon aldehyde), and fructose is a ketohexose (a six-carbon ketone).

Each carbon atom bearing a hydroxyl group (-OH), with the exception of the first and last carbons, are asymmetric, making them stereocenters with two possible configurations each (R or S). Because of this asymmetry, a number of isomers may exist for any given monosaccharide formula. The aldohexose D-glucose, for example, has the formula $(CH_2O)_6$, of which all but two of its six carbons atoms are stereogenic, making D-glucose one of 24=16 possible stereoisomers. In the case of glyceraldehyde, an aldotriose, there is one pair of possible stereoisomers, which are enantiomers and epimers.

1, 3-dihydroxyacetone, the ketose corresponding to the aldose glyceraldehye, is a symmetric molecule with no stereocenters). The assignment of D or L is made according to the orientation of the asymmetric carbon furthest from the carbonyl group: in a standard Fischer projection if the hydroxyl group is on the right the molecule is a D sugar, otherwise it is an L sugar. Because D sugars are biologically far more common, the D is often omitted.

CONFORMATION

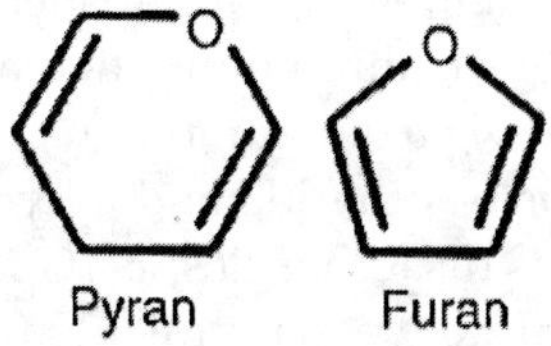

Pyran and furan, after which the pyranose and furanose rings forms of monosaccharides are named.

The aldehyde or ketone group of a straight-chain monosaccharide will react reversibly with a hydroxyl group on a different carbon atom to form a hemiacetal or hemiketal, forming a heterocyclic ring with an oxygen bridge between two carbon atoms. Rings with five and six atoms are called furanose and pyranose forms, respectively, and exist in equilibrium with the straight-chain form.

During the conversion from straight-chain form to cyclic form, the carbon atom containing the carbonyl oxygen, called the anomeric carbon, becomes a chiral centre with two possible configurations: the oxygen atom may take a position either above or below the plane of the ring. The resulting possible pair of stereoisomers are called anomers. In the a anomer, the -OH substituent on the anomeric carbon rests on the opposite side (trans) of the ring from the CH_2OH side branch. The alternative form, in which the CH_2OH substituent and the anomeric hydroxyl are on the same side (cis) of the plane of the ring, is called the B-anomer. Because the ring and straight chain forms readily interconvert, both anomers exist in equilibrium.

USE IN LIVING ORGANISMS

Monosaccharides are the major source of fuel for metabolism, being used both as an energy source (glucose being the most important in nature) and in biosynthesis. When monosaccharides are not needed by cells they are quickly converted into another form, such as polysaccharides.

DISACCHARIDES

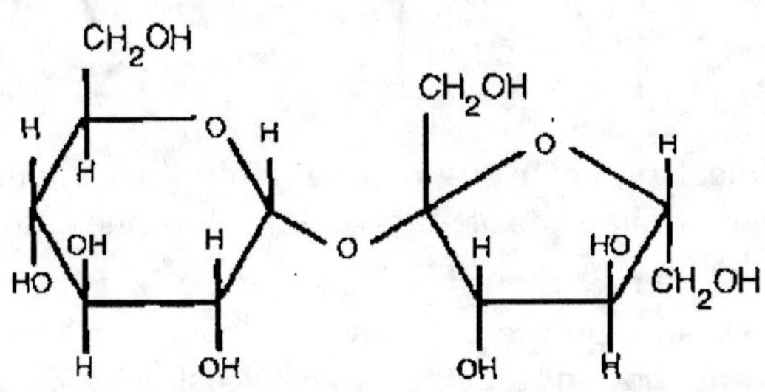

Sucrose, also known as table sugar, is a common disaccharide. It is composed of two monosaccharides: D-glucose (left) and D-fructose (right).

Two joined monosaccharides are called disaccharides and represent the simplest polysaccharides. Examples include sucrose and lactose. They are composed of two monosaccharide units bound together by a covalent bond known as a glycosidic linkage formed via a dehydration reaction, resulting in the loss of a hydrogen atom from one monosaccharide and a hydroxyl group from the other. The formula of unmodified disaccharides is $C_{12}H_{22}O_{11}$. Although there are numerous kinds of disaccharides, a handful of disaccharides are particularly notable.

Sucrose, pictured to the right, is the most abundant disaccharide and the main form in which carbohydrates are transported in plants. It is composed of one D-glucose molecule and one D-fructose molecule. The systematic name for sucrose, O-α-D-glucopyranosyl-(l->2)-D-fructofuranoside, indicates four things:

- Its monosaccharides: glucose and fructose
- Their ring types: glucose is a pyranose, and fructose is a furanose
- How they are linked together: the oxygen on carbon number 1 (Cl) of α-D-glucose is linked to the C2 of D-fructose.
- The -oside suffix indicates that the anomeric carbon of both monosaccharides participates in the glycosidic bond.

Lactose, a disaccharide composed of one D-galactose molecule and one D-glucose molecule, occurs naturally in milk. The systematic name for lactose is O-β-D-galactopyranosyl-(l-->4)-D-glucopyranose. Other notable disaccharides include maltose (two D-glucoses linked α-1, 4) and cellobiose (two D-glucoses linked β-1, 4).

OLIGOSACCHARIDES AND POLYSACCHARIDES

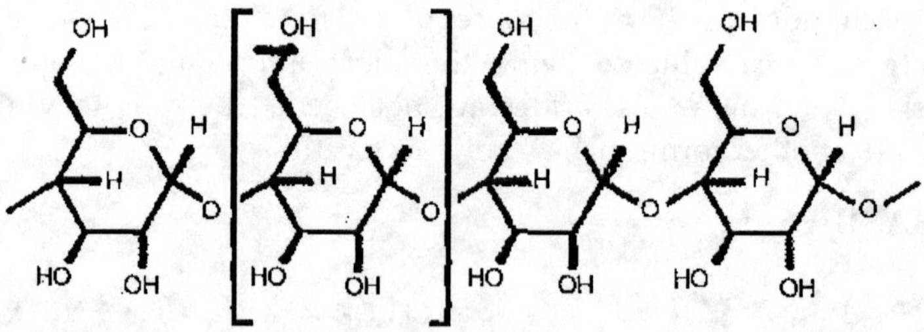

Amylose is a linear polymer of glucose mainly linked with α(l->4) bonds. It can be made of several thousands of glucose units. It is one of the two components of starch, the other being amylopectin.

Oligosaccharides and polysaccharides are composed of longer chains of monosaccharide units bound together by glycosidic bonds. The distinction between the two is based upon the number of monosaccharide units present in the chain. Oligosaccharides typically contain between two and nine monosaccharide units, and polysaccharides contain greater than ten monosaccharide units. Definitions of how large a carbohydrate must be to fall into each category vary according to personal opinion. Oligosaccharides are found as a common form of protein posttranslational modification. Such posttranslational modifications include the Lewis and ABO oligosaccharides responsible for blood group incompatibilities, the alpHa-Gal epitope responsible for hyperacute rejection in xenotransplanation, and O-GlcNAc modifications.

Polysaccharides represent an important class of biological polymers. Their function in living organisms is usually either structure or storage related. Starch is used as a storage polysaccharide in plants, being found in the form of both amylose and the branched amylopectin. In animals, the structurally similar but more densely branched glycogen is used instead. Glycogen's properties allow it to be metabolized more quickly, which suits the active lives of locomotive animals. Cellulose and chitin are examples of structural polysaccharides. Cellulose is used in the cell walls of plants and other organisms, and is claimed to be the most abundant organic molecule on earth. It has a variety of uses including in the paper and textile industry and as a feedstock for the production of rayon (in the viscose process), cellulose acetate, celluloid and nitrocellulose.

Chitin has a similar structure to cellulose but has nitrogen containing side branches, increasing its strength. It is found in arthropod exoskeletons and in the cell walls of some fungi. It has a variety of uses, for example in surgical threads. Other polysaccharides include callose or laminarin, xylan, mannan, fucoidan, and galactomannan.

NUTRITION

Carbohydrates require less water to digest than proteins or fats and are the most common source of energy. Proteins and fat are vital building components for body tissue and cells and are also a source of energy for the body.

Carbohydrates are not essential nutrients: the body can obtain all its energy from protein and fats. The brain cannot burn fat and needs glucose for energy, but the body can make this glucose from protein. Carbohydrates contain 3.75 and proteins 4 kilocalories per gram, respectively, while fats contain 9 kilocalories and alcohol contains 7 kilocalories per gram. Foods that are high in carbohydrates include breads, pastas, beans, potatoes, bran, rice and cereals. Based on evidence for risk of heart disease and obesity, the Institute of Medicine recommends that American and Canadian adults get between 40-65% of dietary energy from carbohydrates.

The Food and Agriculture Organization and World Health Organization jointly recommend that national dietary guidelines set a goal of 55-75% of total energy from carbohydrates, but only 10% should be from Free sugars (their definition of simple carbohydrates). The distinction between "good carbs" and "bad carbs" is an important attribute of low-carbohydrate diets, which promote a reduction in the consumption of grains and starches in favour of protein. The result is a reduction in insulin levels used to metabolize sugars, and an increase in the use of fat for energy through ketosis.

CLASSIFICATION

Dietitians and nutritionists commonly classify carbohydrates as simple (monosaccharides and disaccharides) or complex (oligosaccharides and polysaccharides). Dietary guidelines generally recommend that complex carbohydrates and nutrient-rich simple carbohydrates such as fruit and dairy products make up the bulk of carbohydrate consumption.

The glycemic index and glycemic load systems are popular alternative classification methods which rank carbohydrate-rich foods based on their effect on blood glucose levels. The insulin index is a similar, more recent classification method which ranks foods based on their effects on blood insulin levels. This system assumes that high glycemic index foods and low glycemic index foods can be mixed to make the intake of high glycemic foods more acceptable.

CHEMISTRY OF CARBOHYDRATES

The carbohydrates we find in food fall into one of three groups; mono, di, or polysaccharides. The monosaccharides (and there are normally only three in our diet) are the building units of both disaccharides. and polysaccharides. I will only focus on some few points that are important to understand the fate and function of carbohydrates as an energy source in humans.

GLUCOSE

The central actor in this game is glucose, known also as blood sugar and dextrose. Glucose is a "reducing sugar", that is, it is oxidized by ferric (Fe^{3+}) and cupric (Cu^{2+}) ions yielding reduced metal ions. Glucose also reacts with itself, leading to formation of a "6-ring" (5 carbon atoms and 1 oxygen atom). The really interesting thing with the ring form is that there are two forms of these, a-D-Glucose and β-D-glucose. The first of these has the C-l hydroxyl group on one side of the ring and the 6th carbon in a side chain on the other side. In the beta form of glucose, both of these are on the same side of the ring. Did I hear you say "so what"?

Just wait and you will see that this makes a huge difference in digestibility of polysaccharides made from these compounds.

The carbonyl group in glucose also allows the sugar to interact with many other compounds. In the body this applies especially to various proteins. The level of glucosylated hemoglobin (HbAlc) is used as a measure of blood glucose levels over time. Much of the pathology seen in diabetes is thought to arise from reaction of glucose with the body's proteins.

GALACTOSE

The next of the three monosaccharides we should consider is galactose. This sugar is found in milk and milk products. Galactose is another "reducing sugar" and forms a "6-ring" when dissolved. Galactose is, in fact, almost identical to glucose, the only exception being the position of the hydroxyl group on carbon 4.

Once again, you will soon see that this "minor" difference has large effects on metabolism and can lead to serious symptoms in some newborn babies. As with glucose, galactose undergoes mutarotation and has an alpHa and a beta form.

Only the beta form reacts with glucose to form the disaccharide lactose or milk sugar. Again, this has major effects on digestibility of lactose. We shall soon see that beta-linked disaccharides and polysaccharides are indigestible in humans. Lactose is the only exception to this rule. We have a specific enzyme, lactase, that catalyzes cleaving of lactose. While almost all new-born have this enzyme, most adults do not make lactase and are "milk intolerant".

FRUCTOSE

Fructose is the third of the common sugars found in our diet. As with glucose, it forms ring structures when dissolved. Because the carbonyl group is on carbon two, fructose forms a "5-ring" where 4 carbon atoms and one oxygen atom are involved in the ring structure. Carbons one and six are attached to the central ring and this effects both intestinal uptake and metabolism of fructose. Fructose is a natural but limited component of our diet, being found in limited amounts in many types of fruits and berries. Until about 150 years ago it played a minor role in human nutrition.

Fructose is one of the two sugars that are found in sucrose or "table sugar". It was first in the 1800s that sugar plantations in the West Indies began to produce so much sucrose that this disaccharide became an important part of the western diet. With the increase in "table sugar" or sucrose, came a huge increase in the amount of fructose in our diet. While the consumption of sugar was some few kilograms per year around 1850, the normal annual consumption in most western lands is now around 40-50 kilograms. During the last decade high fructose corn syrup (HFCS) has replaced a good bit of the sugar formerly used in soft drinks, ice cream and cakes. Monomeric fructose and glucose are now used in large quantities in the food industry.

While the use of "sugar", that is common table sugar or sucrose, has decreased in several western countries, total sugar consume has increased greatly. In the USA, sucrose consume is down to 33 kg per person per year. However, total sugar consumption is approximately 60 kg per year. Fructose is not handled as other monosaccharides in our bodies. I will come back to this when I take up digestion of sucrose as well as under a discussion of the metabolism of monosaccharides.

Fructose is sweeter than table sugar (sucrose). For this reason it has been suggested by many "health experts" that one can cut down on the calorie intake by replacing sucrose with fructose. Since fructose is 30-40% sweeter than sucrose, one should be able to eat 30-40% less sugar and get the same sweetening effect. This sounds logical, but is wrong.

Firstly, fructose has a different metabolism than glucose (the other half of sucrose and the same as blood sugar). The enzyme required to initiate fructose metabolism (fructokinase) is only found in the liver. Fructose is not a direct energy source for other tissues. Fructose metabolism is not controlled. It goes very quickly forward unrelated to hepatic energy utilization. This results in an increased synthesis of lipids and increased serum lipoproteins. Fructose is substrate "number one" for gluconeogenesis and gives a rapid rise in serum glucose levels.

That increased sweetness of fructose is a feature of the 5-ring. Now, it just happens that fructose in solution and at higher temperatures (tea and coffee, baking etc.) goes over to a 6-ring form. More correctly, the alpHa and beta furanose forms (5-ring) are in equilibrium with the pyranose (6-ring) forms. The latter dominate in solution and the equilibrium reaction speeds up as the temperature increases. The pyranose forms of fructose are not sweeter than table sugar and one must use just as much fructose as sucrose in warm drinks and baked goods if equal sweetness is desired.

The message should be clear: increased fructose use, either from sucrose or the pure monosaccharide, is associated with increases in blood lipids (increased LDL and decreased levels of HDL) that are associated with cardiovascular disease. Fructose misses its sweetness when warmed and may not help to reduce caloric intake. Sorry, but fructose is not good for you! It's expensive too.

ADSORPTION OF SUGARS IN THE INTESTINE

Adsorption of sugars in healthy persons occurs almost exclusively in the the small intestine. Adsorption is limited to the monosaccharides glucose, fructose and galactose. These "mono-sugars" are large molecules and can only cross cell membranes when a "carrier protein" is present. These are very specific and do not react with and transport other sugars.

Note that uptake of glucose and galactose is through a Na+-coupled symport driven by ATP. This is an active process. Uptake of these sugars is not dependent on a concentration gradient over the intestinal wall. Glucose and galactose are completely "sucked up" from the intestinal lumen and transferred over the mucosal cell to the intestinal circulation.

Fructose is not bound and moved by this active transporter. One of the glucose transport proteins (GLUT5) carries out this job. GLUT5, just as the other members of the glucose transport protein family, is passive. It moves the sugar down a concentration gradient. That is, it requires that the fructose concentration in the intestinal lumen is higher than that in the mucosal cell. Transport from these cells to the circulation must also go "downhill". It is GLUT2, another glucose transport protein that carries fructose as well as glucose and galactose over the basolateral side of the intestinal cell.

Why is the concentration of fructose in the portal blood so low that a concentration gradient is maintained during fructose uptake? The key to this is the liver's GLUT2 and fructokinase. These are very active and fructose is speedily removed from portal blood into the liver and trapped there as fructose-l-pHospHate. This maintains the fructose concentration gradient over the intestinal cells in spite of the rapid absorption of the sugar.

Intestinal inflammation reduces uptake; the enzymes responsible for splitting of disaccharides are produced in the intestinal mucosal cells. Inflammation causes temporary loss of these and a reduced uptake of sugars. Leakage of mono and disaccharides over intestinal membranes can also take place when these are inflamed.

OTHER MONOSACCHARIDES USED AS DIETARY SUPPLEMENTS

There is no uptake mechanism for other monosaccharides in the small intestine. As we shall soon see, lacking or reduced uptake in the small intestine leads to further transport of sugars to the large intestine. There, bacteria overtake digestion, giving gas formation, pain and diarrhea when the transfer of sugar to the large intestine exceeds around 4-5 grams daily. This lack of small intestine uptake for some monosaccharides permits them to be used as "calorie-free" sweetening agents.

SWEET SORBITOL

Sorbitol is a sugar alcohol produced in our bodies by reduction of glucose and and which is further metabolized to fructose. It is readily oxidized and should be a good source of energy. But, it is not taken up in the small intestine since no sorbitol carrier is found here. Therefore, ingested sorbitol can give food a sweet taste without contribution to the energy content of food. Sorbitol is used in chewing gum and some "drops" as a "non-fattening" sweetener. As long as one takes less than those 4-5 grains all is well. More can give "stomach pain" and diarrhea.

RIBOSE

Ribose is an essential part of our metabolic system, being found in pHospHorylated form in, among other things, ATP. The diet supplement industry has presented ribose as a strength and energy giving stuff. In a way, the statement is correct, but the body builds all the ribose-5-pHoshate it needs from glucose. Dietary ribose is completely unnecessary for ATP synthesis. And, ribose is not adsorbed in the small intestine! The label states that the daily dose should be around 4-5 grams, so the producer seems to know that this stuff is not taken up.

A Norwegian weight lifter was disqualified at the Olympic Games in Australia a few years ago. He had taken ribose daily but stopped before the Olympics. He began to lose strength, he said, and began with ribose again Ribose "worked" for him. But why? He was taken for doping with Nandrolon, an anabolic steroid. His ribose was found to be "contaminated" with the stuff. He has sued the producer for this. The bottom line: there is no need for dietary ribose and no intestinal uptake of this sugar!

THE DISACCHARIDES AND THEIR DIGESTION.

All good things come in "threes" when we study carbohydrates. First three monosaccharides, now three disaccharides, each synthesized from two of those simple sugars.

THE BOND

The first thing we must have straight is the nature of the bonds between the sugars in a disaccharide or polysaccharide. These ester bonds are formed through a dehydration of two hydroxyl groups, one on carbon 1 and the other on carbon 4 of either glucose or galactose, or carbon 2 of fructose. If the carbon-1 hydroxyl group is in the beta position (same side of the ring as carbon 6), the resulting bond is a beta 1-4 glucosidic bond. If the hydroxyl group was under the ring (an alpHa hydroxyl group) the resulting bond is called an alpHa 1-4 or alpHa 1-2 glucosidic bond. None of the enzymes that cleave glucosidic bonds in our digestive tract can cleave beta 1-4 linkages. Well, there is ONE exception, lactase, which is specific for the beta 1-4 glucosidic bond in lactose, the sugar found in milk.

SUCROSE

The disaccharide sucrose is composed of glucose and fructose joined by an alpHa 1-2 bond. Sucrose is a relative "newcomer" in our diet. It is actually not especially sweet, and we use large amounts to get the taste we desire in breakfast cereals, soft drinks, cakes and sweets. Just imagine using the same amount of salt in a recipe! Sugar has become a major energy contributor.

A year's consumption of around 50 kilograms of sucrose calculates out to about 500-600 kilocalories per day (our total requirement is around 2000-2500 kcal/day). The problem is that we use sugar as though it was a spice, while we pile on a lot of empty calories in the process! Sucrose is readily cleaved by sucrase, one of the enzymes produced by and bound to the mucosa cells of the small intestine. An important point to remember here is that the activity of sucrase does not normally limit the rate of absorption of the monosaccharides that result from cleavage of sucrose (glucose and fructose). There is ample enzymatic activity to digest the amounts of sucrose usually consumed by adults.

However, intestinal infections or inflammation may well give rise to a temporary loss of mucosa cells with a loss of enzyme activity. Any carbohydrate that reaches the large intestine will serve as a substrate for bacteria there. The resulting gas formation coupled with release of two and three-carbon products from bacterial metabolism can give pain and perhaps diarrhea. We see the same pHenomena in children who eat and drink too many sweets. They simply take in more sucrose than they manage to digest. The resulting tummy ache is the result of bacterial gas formation in the large intestine.

LACTOSE

Lactose (or milk sugar) is a disaccharide composed of galactose and glucose. The enzyme responsible for synthesis of lactose is specific for the beta form of glucose and the result is a beta 1-4 glucosidic bond between the two monosaccharides. Lactose is the only substance with such a bond that can be digested in the human intestine. This requires the presence of lactase, another of the enzymes involved in carbohydrate digestion that is produced and bound to the mucosal cell membrane. In contrast to sucrase and isomaltase, lactase production is restricted and can limit the digestion of lactose. Do you remember Mom who said "you mustn't drink milk when you have a tummy ache"?

She "knew" that lactase production was reduced when the intestinal mucosa was enflamed. The result of drinking milk and other fresh dairy products in the absence of lactase is the transport of the sugar to the large intestine. Bacterial digestion of this leads to gas production. In addition will bacteria produce two and three carbon compounds that increase the osmotic pressure of the intestinal contents, thus retaining water. The result; "explosive diarrhea".

Lactase is usually found in the intestine at birth in all human races. In a global prospective one can state that production begins to decrease at the one to two-year stage and is all but lacking from the age of five years. Lactase-deficient adults can utilize milk products without discomfort if they are cultured with bacteria or yeast before ingestion. Culture milk, sour cream, yogurt and many cheeses are typical products in which microorganisms are allowed to digest lactose for us. Lactose deficiency in adults is quite rare in Scandinavia. Examine this map.

PERCENT OF THE ADULT POPULATION WHO ARE LACTOSE-INTOLERANT

Most of the people in Sweden, Denmark, England, Ireland, Norway (not shown but about 5% intolerant) and northern Germany have lactase activity as adults. In southern Europe the percent of the population that is lactose-intolerant is much larger. Asians, many tribes in Africa, Jews and many others lack lactase as adults.

A THEORETICAL EXPLANATION DISTRIBUTION OF ADULT LACTASE

Glaciers covered Scandinavia during the last ice age. Migration to this largely uninhabited region began about 10,000 years ago as the ice began to retreat. Hunters followed the ice and reindeer and established a growing colony on the newly uncovered land. A mutation leading to life-long production of lactase must have occurred in the Scandinavian population at sometime during these 10,000 years. Otherwise, we would have expected an even distribution of this gene and enzyme in Europe. Where do we find milk-tolerant people in Europe? In the UK, Ireland, Spain and along the rivers of Central Europe; areas of exploitation and commerce during the Viking period.

MALTOSE

Maltose and isomaltose are not major components of our diet but are formed from polysaccharides in the intestine through the action of amylase. These disaccharides are glucose dimers and are digested by maltase and isomaltase that are produced by and bound to the mucosal cell. The glucose formed is then readily adsorbed.

POLYSACCHARIDES

Polysaccharides are the basis of the agriculture-based diet. When we examine the recommendations from the Norwegian Council for Nutrition and most other official agencies, we see that about 55-60% of the energy supplied should come from carbohydrates. Further, that most of these should come from starch and not sugar. That big increase in fructose consumption really appears to lead to CVD and overweight.

Most medical and official dietary recommendations agree with this standard. However, there is one area that I do wish to consider: the differing structures of the digestible and indigestible polysaccharides. We have seen that glucose units (the building stones of the polysaccharides in nature) can be connected with either alpHa 1-4 or beta 1-4 bonds. This is dependent upon the specificity of the enzyme that catalyses formation of these polysaccharides. Some require alpHa glucose, others beta glucose. The bonds formed arrange the connected glucosyl groups at differing angles to each other.

Beta 1-4 linked glucosyl units lie in a single plane while alpHa 1-4 bonded glucosyl groups lie at an angle to each other. This "minor" difference results in entirely dissimilar 3-dimensional structures. Beta 1-4 linked glucosyl groups form sheets that pack together to form the rigid structure we know as cellulose. This is the major polysaccharide of grass, leaves and trees and is said to include around 50% of all biological carbon found on our planet. Digestion of cellulose is entirely dependent upon intestinal flora in herbivores. The anatomy of our digestive system precludes the presence and function of these organisms. Cellulose is, however, of importance in human nutrition in that fibre (good old indigestible cellulose) is an essential part of our diet, giving bulk to our food and promoting intestinal motility.

The $\alpha(1 \rightarrow 4)$ glucosyl groups in these polysaccharide form structures with thousands of glucosyl groups bound together in a spiral. These are found packed together in granules. The granules bind water and the enzymes necessary for their synthesis and catabolism, providing a compact system for very rapid rates of synthesis and breakdown of starch and glycogen.

BLOOD SUGAR VALUES, NORMAL AND DIABETIC

The upper limit for normal levels of blood glucose is currently set at less than 5.6 mmol/1 while the reference level 2 hours after a glucose load is less than 7.8 mmol/l. Note that interpretation of the data is "temporary". Glucose levels swing somewhat from day to day and from individual to individual. Therefore, if high glucose levels are found, it is advised that measurement should be repeated on the following three days. Only then should hyperglycemia be noted as an enduring symptom and treatment started.

Fasting glucose levels between 5.6 mmol/1 and 7.0 mmol/l indicate impaired glucose tolerance (IGT) while higher levels signify diabetes. IGT can be viewed as "pre-diabetes" as most individuals in this group develop diabetes type 2 with time.

In 2003 an expert group suggested reducing the cut point for impaired glucose tolerance from 6.1 mmoles/l to 5.6 mmoles/l. This is expected to increase the number of individuals diagnosed as prediabetics by about 20%. Many of these people have a large risk for development of diabetes type 2 later in life. Therefore, they should be advised to revise their lifestyle and work for reduction of their hyperglycemia.

Fasting blood glucose levels are proposed as the leading indication of the metabolic state. This is a rapid, inexpensive and accurate measure of the individual's metabolic situation. The committee suggested that fasting blood sugar be determined every 3rd year after reaching 45 years of age. A glucose tolerance test is deemed unnecessary as long as fasting glucose does not exceed 5.6 mmoles/liter.

WHATS WRONG WITH HAVING A LITTLE EXTRA BLOOD SUGAR?

GLUCOSE INTOLERANCE AND THE DEVELOPMENT OF DISEASE

The suggested blood glucose standards are based upon statistical analysis of the correlation between of blood glucose levels and the development of pathological states. Among these is retinopathy, which can lead to blindness. One might have expected a linear relationship between the level of blood sugar and the development of glucose-related diseases.

Data from three independent studies are shown in which elderly Americans, Pima Indians and Egyptians were observed. These groups are known to be especially prone to development of diabetes type 2. In all cases, an abrupt increase in the frequency of retinopathy was noted when blood sugar levels exceeded 6.0-6.1 mmol/l. This toxic effect of glucose may result from sorbitol formation or follow glycation of proteins, a non-enzymatic coupling of glucose to proteins.

CLASSIFICATION OF DIABETES MELLITUS

Most diabetic patients fall into one of two classifications, diabetes type 1 and diabetes type 2. In the Asia, USA and Europe about 80% of all diabetic patients have the type 2 variant

Classification of Diabetes Mellitus

(American Diabetes Association 1998)

- Type 1: B-cell destruction usually leading to absolute insulin deficiency.
- Type 2: Variation from insulin resistance and increased insulin levels to a dominant defect in insulin secretion with insulin resistance.
- >40 well-defined types of diabetes.
- Gestational diabetes: Any degree of glucose intolerance first noted in pregnancy.

Patients with diabetes type 1 must receive insulin to survive. Patients with type 2 diabetes and glucose intolerance are best treated by learning to control their own metabolism through proper diet and motion. In addition, many use metformin to reduce the liver's glucose production.

Specific treatment of the detailed cause of diabetes type 2 is seldom possible. It would be of great advantage to have exact knowledge of the origin of glucose intolerance and diabetes type 2 in each patient. More than 40 differing causes have been found in some very few cases. These mainly involve enzyme mutations. This is an active research area and may, with time, lead to advances in treatment.

Gestational diabetes is a passing hyperglycemia in pregnancy. This is a genetically determined condition and is handled without insulin treatment. Women who acquire gestational diabetes often develop diabetes type 2 later in life.

DIABETES TYPE 1

Insulin was first identified in pancreatic extracts by Banting and Best in 1922. The first patient who received insulin was 14 year old Leonard Thompson, also in 1922. Diabetes type 1 was a fatal disease prior to Banting and Best's work. Treatment of diabetes type 1 with insulin is still not economically possible for many patients in underdeveloped countries. Diabetes type 1 results from an autoimmune destruction of the β-cells of the pancreas. It has been most commonly accepted that this is a genetically controlled process associated with HLA alleles. However there is a agedependent spreading of the disease and this has changed with time. Increasing evidence points to a combination of factors leading to beta cell destruction. The disease most commonly develops before 20 years of age. However, incidence (cases/100,000) of the disease has changed in the latter half of the 20 th century. Patients develop diabetes at an earlier age and the frequency has increased. This implies that there are other factors than the genetic which lead to development of diabetes type 1.

AUTOIMMUNE ATTACK ON THE LANGERHANS ISLETS

It has been commonly assumed that antibodies and T cells specifically attacked beta cells, leading to insulin deficiency and destruction. However, in a recent publication, Shawn Winer et al have shown that Schwann cells which surround the Langerhans islet are autoimmune targeted and destroyed before the autoimmune attack on beta cells and loss of beta cell activity. It appears that antigen-presenting cells move from islets to the pancreatic lympH mode where cell-specific antigens are produced.

These attack Schwann cells surrounding the islets, destroying the Schwann cell capsule which surrounds islets. Beta cell destruction follows. Diabetes type 1 patients are today completely dependent upon injections of insulin to sustain life. Current research is aimed at developing oral agents capable of activating the insulin-signal pathway downstream of the insulin receptor. In this way the need for injected insulin may perhaps be eliminated in the future.

WHY DON'T WE REPLACE LOST CELLS?

One might think that we could replace destroyed? -cells and restore insulin secretion and control of metabolism. After all, blood cells, skin and intestine do not disappear after loss of tissue. This was the question a group of researchers at Harvard put forward in a recent study. They investigated generation of β-cells in mice and looked for formation of new pancreatic islets in adult animals. Surprisingly, they found that no new islets could be identified. In adult mice, β-cells are formed only by self-duplication and appeared not to arise from stem cells. The permanent loss of β-cells seen in diabetes type 1 may, therefore, occur because only preexisting β-cells give rise to new insulin-producing cells in the adult. The loss of insulin producing β-cells seen with time in diabetes type 2 may also develop through this mechanism.

IS REPLACEMENT THERAPY POSSIBLE?

A fascinating approach to diabetes type 1 therapy would be to replace pancreatic beta-cells with new cells produced in the laboratory. Is this possible? Can we produce human beta cells in vitro? Gershengorn et al present data showing conversion of human pancreatic fibroblast-like cells to insulin producing cells. Cell replacement therapy just might be possible, but it remains to be seem whether or not fibroblasts in vivo can give rise to viable insulin-producing cells.

DIABETES TYPE 2

Diabetes mellitus type 2 arises from a reduced response of target tissues to insulin (so-called insulin resistance). Insulin levels are often quite high in early type 2 diabetes. The etiology of insulin resistance remains unclear, although many hypotheses have been set forth. I will go into these later, but there is a very clear association between overweight (BMI 25-30) or obesity (BMI>30), the metabolic syndrome and development of diabetes type 2. A genetic factor is also clearly involved. Many ethnic groups are predisposed for the disease.

Diabetes type 2, earlier called "maturity -onset diabetes", typically develops over a period of many years. It usually begins with overweight, goes through a period of glucose intolerance, and ends with the full-blown diabetic state. Obesity seems to be the most common factor initiating development of diabetes type 2. The patient goes through a period of increasing insulin resistance during which the body compensates by producing higher levels of insulin after meals. Usually, the patient is unaware of his or her situation. Following a period of many years, the pancreatic β-cells no longer manage to continue their enhanced insulin secretion. Insulin levels fall in the face of continuing insulin resistance, resulting in increased fasting blood glucose levels and markedly increased post-meals levels of plasma glucose. It has been suggested that the insulin-producing β-cells are subject to attack even before this stage.

It should be noted that the global wave of obesity we now experience applies to young children too. Social factors (fear of kidnapping, parents who drive kids to school and afterschool activities, passive TV and PC use, etc.) contribute to reduced pHysical activity and obesity in children. Simultaneously, energy intake is often increased through an ever-increasing consumption of fast foods, soda pop, sugar coated breakfast food, snacks and confection.

The decreased pHysical activity coupled to increased food intake must lead to pronounced weight gain. These energy rich foods appear not to function normally in the appetite-controlling mechanism. The reason for this is not yet clear but is the subject of much research. Diabetes type 2 in young children (formerly limited to adults) is now a serious and growing problem. Control of the long-term symptoms of diabetes type 2 will be a difficult and very expensive problem in the coming decades.

A SIMPLISTIC APPROACH TO WEIGHT REGULATION

There are very many factors involved in normal weight regulation. However, the end result is the consequence of the balance between the amount of energy in the ingested food and the amount of energy used. Hormone levels, altered hormone receptors, mutated metabolic enzymes; all of these do influence appetite and energy use. However, the bottom line is, you are what you eat (and do not burn). The activity levels shown here correspond to those that are normal in a modern society. A young male student uses most of the energy in his food to support normal body functions (basal metabolism or BMR). Only about 20% of the consumed energy go to support pHysical activity.

If one leads a more active life (cycling to work, using stairs instead of the elevator, some sports after studying), the total daily energy use increases to about that level that the average person used for 100-150 years ago. It is estimated that, at that time, both men and women used about 3000 kcal/day for the usual daily chores. Note that at this activity level pHysical activity still accounts for not more than 25-30% of the total energy used daily.

A heavy activity level, as judged by our modern standards, will increase energy use to around 3500 kcalories/day. By the way, lumberjacks using axes and hand saws are said to have used about 5000 kcalories per day! Most experts believe that a sedentary or passive life style is the best explanation for the trend to overweight. That is, weight gain is more closely related to energy use than energy intake. What has changed in the past decade?

A recent editorial in Mayo Clinical Proceedings listed some examples of sedentary and active ways of doing things. Many of the recent advances (read labour-saving) in our daily environment reduce energy utilization. If you make some assumptions you can calculate the approximate difference in total energy use between the "easy and hard" ways of doing common tasks. This lies somewhere around 10,000-15,000 kcal every month or two. That is just about the equivalent of 1-1,5 kilograms of body fat. Few of us take this into account when we plan our meals! Doing things the active way, a partial switch to a plant-based diet with no more than 30-35 calorie % fat and moderate daily motion for between 30-60 minutes should help keep the fat away for many people.

We all have heard that "our western way of life" leads to the pHysical downfall that we have experienced during the past 10-20 years. The fact is that obesity is a global pHenomena linked more to pHysical activity than race and culture. WHO has stated "At the other end of the malnutrition scale, obesity is one of today's most blatantly visible - yet most neglected - public health problems.

Paradoxically coexisting with malnutrition, an escalating global epidemic of overweight and obesity - "globesity" - is taking over many parts of the world. If immediate action is not taken, millions will suffer from an array of serious health disorders". Let us look at rates of obesity in a global perspective. We have most data from USA and England, but also some data from South America and the Pacific area. About 25-30% of the populations of USA and England have now a BMI of 30 or more. We see the same trend in a little island in the Indian Ocean (Mauritius) and the trend in Brazil is not much better. The most striking figures come from the Pacific region. Here, around 70% of the adults on some islands are obese. The prospects facing health officials and pHysicians are grim.

WHY DON'T WE EAT LESS?

The amount of food we consume each day is regulated by what we call appetite; the desire to eat. The hypothalamus is the central coordinator of signals relating to hunger and satiety. The genetic basis for this very complicated process has developed over a period of millions of years. Rates of both basal and total metabolism are given. Both BMR and total metabolism seem to have been more or less constant from earliest prehistoric times to those seen in now-living "primitive" people (groups who base their existence on hunting and gathering). An energy expenditure of around 200 kJ/kg per day calculates out to about 3200 kcal/day for a 70 kg person.

About 150 years ago most people used around 3000 kcal per day to come through a day's chores. Work was done manually. She washed, cleaned and raised children without a car, a washing machine, vacuum cleaner, or a dish washer. Food was prepared from raw products on a wood stove. He built and worked on the farm without tractors, harvesters and chain saws or in a factory or mine with heavy manual tasks. Modern life goes back no further than to the beginning of the 20th century. Modern or "post-modern" styles of life have existed for no more than 100 years for people in industrial societies and far less for those in developing countries. The current imbalance between intake of food and expenditure of energy may well have followed the introduction of powered labour-saving devices, during this short period. Genetic adaptation to changes in environment occur slowly through any generations. Stable genetic adaptation to new environments in whole populations do not occur "overnight".

Now, the fact is that the frequency of obesity has not increased linearly during the past century but quite rapidly during the past decade. Remember, obesity is defined as having a BMI of 30 or more. In 1991, only 5 states had a population in which 15% or more were obese. By 2000, all but one state could boast of this. In most of these over 20% of the population had a BMI over 30. Overweight is probably as large a threat to health as hunger in today's strange world. One of my colleagues was recently in India.

He reported that the "man on the street" in southern India was as thin as we in the West usually imagine him to be. However, obesity was very frequent at Indian airports. Good economy, easily attained food, a car and other labour-saving devices lead to weight gain, metabolic syndrome and poor health.

SUGAR AND HEALTH

The yearly intake of sugar has risen from 4-5 kilogram in the 1850s to about 40-50 kilos in the 1990's. Forty kilos equal about 400 kcal/day or around 15-20% of the energy needed to maintain normal activity. Much of this sugar comes from soft drinks. These contain approximately 30-40 sugar bits per liter! Sugar is used as a "spice", giving a proper taste without a thought about its energy content.

Now, the fact is that many individuals consume much more sugar than the alleged 40 kilograms yearly. That figure often includes only "table sugar" or sucrose. In reality, soft drinks and cakes contain a lot of monomeric sugar (fructose and glucose) made from corn syrup. While the officially recognized sugar consumption in the USA in 1997 was 33 kg/person annually, the actual total sugar consumption was over 60 kg/person/year.

Production of high fructose corn syrup began in the '70s. While it is quite true that sucrose intake was reduced in the following years, the total sugar consume increased dramatically. Total glucose and fructose, derived from sucrose and corn syrup increased by almost 30% during these 27 years. The high sugar consume is a major factor in increased caloric intake. Perhaps even more important is the very rapid and metabolically uncontrolled metabolism of fructose. Both of these factors contribute to hepatic triglyceride production and augmented plasma triglyceride levels. There is a close correlation between plasma triglycerides and arthrosclerosis and development of cardiovascular disease. In the USA, agricultural subsidies have increased production and consumption of sweet and fat food.

METABOLIC SYNDROME OR SYNDROME X

The relationship between diet, metabolism and poor health expressed by the term "metabolic syndrome" was noted simultaneously by cardiologists and endocrinologists about 10-15 years ago. While the numbers of CVD patients was decreasing, an increasing percent of new cardiac patients were found to be overweight, have increased blood sugar, dyslipidemia and hypertension. These symptoms were also noted in the ever-increasing number of patients with diabetes type 2. This condition was called "Syndrome X" or metabolic syndrome. There are several definitions of this syndrome. The ATP III has defined metabolic syndrome shown in the next figure (reference values in italics). Those filling 3 of the 5 criteria have metabolic syndrome. These criteria are objective, being based solely on observations seen in many thousand patients. Overweight is characteristic of most of these, especially those with an accumulation of fat surrounding the intestines (central obesity). It has been suggested that this is because of the relatively large metabolic activity of this tissue. A more recent suggestion is that abdominal fat produces cortisol that may influence metabolism.

A newly discovered peptide hormone, omentin, which increases food consumption is also produced here. Many of these individuals show high levels of serum total triglycerides and this is often observed in overweight. High density lipoprotein levels (HDL) are often lower than normal. Interestingly, LDL levels do not correlate well with the pathological changes seen in metabolic syndrome. However, a shift to large lipid-rich LDL particles is often seem. Hypertension is common in metabolic syndrome as is hyperglycemia. Dependent upon the level of blood glucose, these patients can be classified as having impaired glucose tolerance (IGT) or diabetes type 2.

The scope of the metabolic syndrome problem is enormous. Weight gain is common among older persons with a gradually increasing BMI from 30-40 years of age. This correlates well with development of metabolic syndrome as shown in the next figure from the USA. This data is not current. The percent of the population with metabolic syndrome and the subsequent health problems is much higher today. There is now a growing tendency for children, both in the USA and Europe, to develop glucose intolerance (IGT) and diabetes 2 parallel with obesity.

WEIGHT REDUCTION AND DIABETES TYPE 2

The importance of overweight in the development of metabolic syndrome and diabetes type 2 cannot be overrated. Simply dieting and reducing weight can be sufficient to improve a patient's situation. This figure from Medscape/Diabetes Care shows the effect of weight loss on the response to a 75g glucose load.

After 12 weeks of dieting, fasting glucose levels were reduced as were the plateau levels following a glucose load. We know from this and many other studies that the difference noted is dependent upon increased sensitivity to insulin. In other words, weight loss often leads to decreased glucose resistance and a normalized metabolism.

GLOBAL PERSPECTIVES OF THE INCIDENCE OF DIABETES

Predictions of the global frequency of metabolic syndrome, insulin resistance and diabetes are frightening. The consequences of urbanization and overweight are much more threatening that the AIDS/HIV epidemic. Most of the increase in diabetes will be that of type 2. Overweight with associated insulin resistance and metabolic syndrome will lead to an almost 50% increase in the number of individuals suffering from diabetes within the current decade. Over 200 million persons will be involved.

Most of this expansion will occur in developing countries where health services will be unable to cope with the coming situation. Without indispensable guidance and treatment, these people will develop coronary and vascular disease, chronic kidney disorders, hypertension and the many other complications of diabetes. Predictions for the middle of the century indicate that at least 300 million persons will develop diabetes type 2. Genetic background is a strong component in the development of type 2 diabetes. People of western European origin are less sensitive to the effects of overweight than non-Europeans. In the USA, blacks and Latinos have a higher frequency of diabetes type 2 than whites. Asians, especially Chinese and Japanese, develop metabolic syndrome and diabetes type 2 at lower body weights than whites. Thus, these groups develop diabetes 2 in the BMI range from 25-30.

Prevalence of diabetes is relatively low in Europeans and native people living in their accustomed environment. However, native people who convert to a western diet or urban life present a markedly increased incidence type 2 diabetes. Genetic variation can explain differing frequencies of diabetes development in diverse racial groups. However, development of type 2 diabetes among urbanized groups must be a response to the environment. Genes do not change so quickly!

One explanation for this has been the development of a set of "thrifty genes", a genetic adaptation to a poor milieu with a meager diet. A switch to an urban diet and activity level is suggested to rapidly lead to obesity and associated illnesses in such people.

INSULIN RESISTANCE, THE KEY TO DIABETES TYPE 2

The overweight persons depicted here were on the way towards developing glucose intolerance, but still had control over their blood sugar levels. We can clearly see that obesity is accompanied by insulin resistance. The obese persons produced more insulin after meals than controls to counteract this. In this way they can counteract their insulin resistance and regulate glucose uptake and hepatic glucose production.

Diabetes type 2 develops in response to a loss of reactivity to insulin at target organs. To counter this, the pancreatic? - cells produce more insulin to counter the decreased target organ sensitivity. Initially, metabolic control will be normalized through this mechanism. However, with time, the individual will develop impaired glucose tolerance and diabetes type 2.

The panel on the left shows that glucose disposal (glucose taken up from the blood) is reduced in type 2 diabetes. This indicates that muscle and adipose tissue take up less glucose in these patients than control persons, leading to hyperglycemia after meal.

HOW DOES INSULIN RESISTANCE DEVELOP?

The key to this problem seems to lie in the interaction between adipose tissue and skeletal muscle. In "the old days", we looked on fat as a relatively inactive, not especially attractive part of our bodies. We now know that adipose tissue is metabolically quite active and is one of the body's most important endocrine organs. No less than 14 peptide hormones have been found in adipose tissue. In the past few years three of these, leptin, adiponectin and resistin have received most attention.

Another central function of insulin is inhibition of hepatic gluconeogenesis. Remember that it is the balance between glucagon and catecholamines on one side, and insulin the other that directs hepatic gluconeogenesis and release. Glucagon and the catecholamines stimulate glucose production and glycogenolysis when blood glucose levels fall: insulin inhibits gluconeogenesis when there are adequate levels of glucose in the blood. In the right-hand panel we can see that hepatic glucose production is less sensitive to insulin in diabetes type 2. In spite of high blood sugar levels, the liver produces glucose and releases it to the circulation. People suffering from diabetes synthesize glucose in spite of the fact that blood glucose levels are elevated! Therefore, treatment with a glucose synthesis inhibitor, metformin, is often started in addition to adjusting to a new life style.

Leptin seems to have many actions; it is involved in thermoregulation, bone metabolism, fatigue and especially appetite regulation together with insulin and blood sugar. Current opinion seems to be that leptin is a signal of reduced energy and stimulates hunger as a result of reduced release from adipocytes. Release of leptin appears to be triggered by the rise in insulin levels seen after a meal. Together, insulin and leptin act centrally to reduce hunger.

Resistin has been implicated in regulation of insulin sensitivity (and insulin resistance) in muscle and adipose tissue. Adiponectin is also possibly involved in regulation of target-cell's sensitivity to insulin. One can conclude that adipose tissue is actively involved in regulation of energy homeostasis through two major functions:

- As a storage tissue for excess nutrients and as a source of energy reserves to all of the body's organs.
- As an endocrine organ involved in regulation of appetite and insulin sensitivity.

CENTRAL CONTROL OF APPETITE

Those of us who studied medical biochemistry a few decades ago learned that appetite was controlled centrally through glucose-sensitive cells in the brain or hypothalamus. Indeed, this is still a "fact". However, we now know that many other factors are involved. Initially, insulin and recently leptin were seem as the major elements controlling appetite. The levels of these are affected by the food we eat, our activity level and our fat stores. I have earlier empHasized the balance between food consumption and pHysical activity.

These factors define "energy balance". While we really are not aware of regulation of appetite, most people managed to hold this balance and a more or less constant body weight before all of our "labour-saving" devices became common. This implies that a very well regulated system must exist. The combined effects of insulin and leptin in the central nervous system appear to fine-tune our urge to eat, leading to a decreased food intake at weight gain and increased appetite following pHysical activity or weight loss. More recently, several additional peptide hormones that couple the digestive system to appetite regulation have been identified. These are ghrelin, released from the stomach and the intestinal hormone PYY3-36. These hormones interact with insulin and leptin sensitive neural cells in the brain. These produce an effect through interaction with hypothalamic neurons producing neuropeptide Y (NPY) and the Agouti-related protein (AgRP) (the GO system) or by activating the melanocortin system: (the STOP system). The balance between "stop and go" signals effects neurons that regulate the sense of hunger and appetite. Insulin and leptin are thought to regulate long-term effects and body weight, while the hormones derived from the digestive system appear to regulate meal-eating on a short-term basis.

WHAT "REALLY" CONTROLS ENERGY METABOLISM AND APPETITE?

It has become increasingly evident that adipose tissue is one of the major endocrine organs in the body. Agonists which bind to receptors in fat tissue are now used to regulate metabolism in type 2 diabetic patients. What are these receptors and what functions do they have?

Peroxisomes are cytosolic vesicles which resemble lysosomes. However, they possess an active oxygen-linked redox system which uses FAD to oxidize substrates. The $FADH_2$ is than oxidized by molecular oxygen, forming H_2O_2 which is normally destroyed within the peroxisome by catalase. The organelles are very active in the metabolism of long-chain fatty acids. A family of nuclear receptors which bind steroids that induce peroxisomal production was found in the early 1990's. These receptors increase "peroxisomal proliferation", and were given the name "peroxisomal proliferator-activated receptors or PPARs.

A series of agonists were then found that bound to PPARs and which indicated that there were several varieties of these receptors in various organs. These are known as PPAR-α, PPAR-γ, and PPAR-δ. PPAR-α is found mainly in the liver, PPAR-γ in adipose tissue and PPAR-δ is expressed in many of the body's tissues but has low activity in the liver. All of these bind dietary fatty acids and their metabolites. Evans and coworkers suggest that the PPARs provide the basis for coordination of metabolism in our various tissues and, in fact, ultimate integration of the body's metabolism.

SOLUBLE FIBRE

There are many other carbohydrates found in varying amounts in our diet, many of which can either be digested or absorbed in the small intestine. These migrate further to the large intestine where they serve as substrates for the intestinal flora. We have all observed the effects of eating large portions of pea soup, baked beans and other legumes. These have a variety of mono-, di- and trisaccharides that cannot be digested in humans.

They are sometimes called antinutrients because on their tendency to lead to gastrointestinal disturbances. A variety of mono, di and trisaccharides that are not digested by our intestinal enzymes but are substrates for the bacterial flora of the large intestine. As in the case of lactose intolerance, the metabolism of these compounds leads to formation of gas and varying degrees of discomfort. Note especially pectin (found in apples) and raffinose (found in legumes-beans). Most of us have experienced the effects of these compounds.

CHAPTER 5

Vegetarian Diets

Is a vegetarian diet healthier than a diet that includes meat? Many believe that it is, but others contend that a vegetarian diet lacks sufficient vitamins and nutrients. A number of people also advocate a vegetarian diet for other reasons such as the environment and animal rights. In the past, some people stayed away from pork, a source of protein, because of its high fat content. That is no longer true. The National Pork Producers Council notes that pork today is leaner and lower in fat, calories, and cholesterol than it was 20 years ago. All of the loin cuts of pork-such as tenderloin and loin chops- "provide less than 180 calories and nine grams of fat per three-ounce serving. " The council says that pork contains about a third less cholesterol than chicken and that there is only a slight difference in the amount of fat between pork and chicken. The National Chicken Council states that fat content breast meat is the part of the chicken lowest in fat, but leg meat is very competitive with other types of meat and poultry. Three ounces of baked chicken drumstick with the skin removed has less total fat than the same amount of sirloin steak, beef tenderloin, pork chop, or salmon. The calories and fat content of chicken compare to other animal protein sources per three-ounce boneless, cooked portion:

Nevertheless, a growing number of Americans are deciding to eliminate animal products from their diets. A similar statistic was obtained for poultry, and 9% said that they never eat fish or seafood. For dairy products, eggs, and honey, those statistics were 3.7%, 6.7%, and 15.4%, respectively. That means that millions of Americans follow some form of vegetarian diet.

There are several types of vegetarians. Lacto-ovo vegetarians eat dairy and egg products in addition to fruits and vegetables; ovo vegetarians eat egg products; lacto vegetarians eat dairy products; and vegans (pronounced VEE-gn) refrain from all animal products including dairy, eggs, and honey. "Beyond modifying their diet, many vegans refuse to consume any animal products. That means they do not wear leather or consume anything else derived from animal source".

Type of Mest	Calories	Total Fat (g)	Saturated Fat (g)	Cholesterol (mg)	Protein (g)
Filet of sole, baked	100	1.5	0.3	60	20
Chicken breast, no skin, baked	120	1.5	0.5	70	24
Chicken drumstick, no skin, baked	130	4.0	1.0	80	23
Salmon, baked	150	7.0	1.5	55	21
Chicken breast, with skin, baked	170	7.0	2.0	70	25
Chicken drumstick, with skin, baked	180	9.0	3.0	75	23

People give many reasons to eliminate animal products from the diet. These include "health, ecological and religious concerns, dislike of meat, compassion for animals, belief in non-violence, and economics.

People often become vegetarian for one reason, be it health, religion, or animal rights, and later adopt some of the other reasons as well". In essence, the article contends that vegetarians are healthier than nonvegetarians and have longer lives. "Vegetarians live about seven years longer, and vegans... about 15 years longer than meat eaters." Vegetarians save money. "Replacing meat, chicken and fish with vegetables and fruits is estimated to cut food bills by an average of $4,000 a year." They also spare the lives of untold numbers of animals. "Each day, 22 million animals are slaughtered to support the American appetite for meat". In Vegan Nutrition: Pure and Simple, Michael Klaper outlines some of the reasons that people may chose to follow a vegetarian diet. "We now live in a world of Salmonella-tainted chickens, Listeria-covered cheese, and beefburgers laced with estrogenic hormones and residues of potent antibiotics.

These are very good reasons why people would want to obtain their daily nourishment without pouring fatty and contaminated meat and dairy products through their bloodstream every few hours.

Many people are concerned about the high incidence of health threats that are linked to improper diet, like heart attacks, strokes, birth defects, breast, prostate and other cancers, high blood pressure, diabetes and other diseases. They are seeking more wholesome food alternatives for themselves and their families. " He adds that eating meat products supports the violence that is part of killing animals for food, and it negatively impacts the environment. Humans have no nutritional requirement to eat any form of animal food or milk from cows. The following are some of the long-term health reasons for eating a plant-based diet:

- A vegetarian diet lowers blood pressure in those people who are prone to high blood pressure.

- Vegetarians tend to have lower cholesterol levels.
- After controlling for socioeconomic status, body-mass index, and smoking, vegetarians have cancer rates that are 20 to 50% below population averages.
- When compared to the general population, vegetarians are far less likely to be obese.
- Plant-based diets tend to support later menarche. That finding has been correlated with a reduced risk for breast cancer.

Vegetarians consume a plant-based diet. What does that mean? "Vegetarians eat a wide range of foods that come from the soil, including fruits, vegetables, grains and grain products, nuts, seeds and legumes, and dried beans and peas". Does such a diet have sufficient protein? Havala writes, "as long as you are eating enough wholesome foods to meet your calories needs and are getting a reasonable mix of vegetables, grains, and legumes over the course of the day, you should not have any problems getting enough protein or the appropriate mix of amino acids. At the same time, you will probably avoid the excessive amounts of protein that can be typical of the traditional Western-style diet".

HISTORY OF VEGETARIANISM

Vegetarianism is not a new concept. It has existed throughout history in cultures around the world. "Many of the world's greatest pHilosopHers and intellects refused meat in times when such a choice was contrary to the dictates of the ruling class. In the west, the first of these was the Greek pHilosopHer Pythagoras, who lived 2,500 years ago and is often considered 'the father of vegetarianism.' Until the late 19th century, when the word 'vegetarian' was coined, people who lived on a meatless diet were referred to as 'Pythagoreans'".

Born around 580 B.C., Pythagoras was multitalented. In addition to discoveries in mathematics and geometry and beliefs in planetary motion and the rotation of the Earth around the sun, "he also founded a society that pursued wisdom, believed in the transmigration of souls and practiced meditation. Among the Pythagoreans, materialism and meat- eating were taboo". Since Pythagoras, there have been a countless number of vegetarians, including Plato, Socrates, Newton, Leonardo de Vinci, Mahatma Gandhi, Albert Schweitzer, Albert Einstein, Paul and Linda McCartney, Richard Gere, Spider-man star Tobey Maguire, and Kim Basinger. It should be noted that until the twentieth century, only the rich had the wherewithal to eat meat on a regular basis. "Even in the 18th and 19th centuries, well-to-do families of many European countries such as France and Holland could seldom afford to eat meat more than once a week.

The truth is that the vast majority of the world's population has existed on a near-vegetarian diet and still does, though not necessarily by choice".

Donald Watson founded the vegan segment of the vegetarian movement in England in 1944. It "called for the abolition not only of all foods of animal origin but of all animal based commodities as well. Furthermore, it empHasized the moral, spiritual, social, health, and economic advantages of living by humane principles".

During the late 1960s and early 1970s, the vegetarian movement gained momentum, especially among counterculture groups that advocated ecology and natural living. In 1971, Diet for a Small Planet, by Frances Moore Lappe, was published. Her arguments for the environment and fighting world hunger attracted many people to vegetarianism.

EFFECTS OF VEGETARIAN DIETS

During the 1990s, increasing numbers of health professionals saw the value of vegetarian diets. Shortly before his death in 1998 at the age of 94, Benjamin Spock, probably the best-known pediatrician in the United States, advocated vegetarian diets for children. Apparently Spock had become a vegetarian in 1991. According to his wife, that change of diet "greatly improved his health and enabled him to complete the revision of his world-famous book".

In the seventh edition of Dr. Spock's Baby and Child Care, Spock and Steven J. Parker wrote, "A vegetarian-based diet for children is generally more healthful than a diet containing the cholesterol, saturated fat, and excessive protein found in meat and dairy products. " But this type of diet-derived from many leafy green vegetables, fruits, whole grains, and bean products-should not necessarily be low in calories. "Studies have shown that a well-balanced vegetarian diet had many advantages and does not interfere with a child's growth and development. A reliable vitamin B12 source such as a children's vitamin or cereal or soy milk fortified with B12 is recommended for vegetarians".

In Pregnancy, Children, and the Vegan Diet, Michael Klaper recommends a vegan diet for children. He claims that animal fats have been linked to clogged arteries in children as young as 9 months old. Fat in children's unbalanced diets is the basis for heart attacks and strokes when they are adults and sometimes in childhood. "A high-fat diet also promotes higher levels of sex and growth hormones that circulate in the body. The effects of these higher hormone levels are of great concern, as growth and maturation are crowded into disturbingly early years".

Are vegetarian diets always beneficial? Do they provide all the essential vitamins and nutrients that people require? First, it should be noted that not all vegetarian diets are low in fat. Whole dairy products, such as cheese, milk, cream, and butter, are high in saturated fat, which is unhealthful.

When consumed frequently, they may negate any of the advantages of a vegetarian diet. Though nuts, seeds, mono-unsaturated oils, avocados, and tofu-foods often eaten on a vegetarian diet-are also higher in fat and calories, they contain healthier fats. Numerous studies indicate that a low-fat vegetarian diet may be helpful for people with a number of medical conditions. A short-term examination of 500 men and women published in the American Journal of Clinical Nutrition found that "a strict, very-low-fat vegetarian diet free from all animal products combined with lifestyle changes that include exercise and weight loss is an effective way to lower serum cholesterol and blood pressure". Women with menstrual cramps and premenstrual syndrome (PMS) have obtained relief from a low-fat, vegetarian diet. Researchers at Georgetown University in Washington, D.C., followed 33 women with moderate to severe symptoms. "The women shunned meat, fish, dairy, and eggs, as well as butter, oils, and even, peanut butter."

Their diets had only about 10% of the calories from fat, "the lowest healthy limit". For many of the women, the results were quite dramatic. "They experienced less intense and less frequent menstrual pain and also noted increased energy, decreased water retention and fewer concentration problems. Some even lost weight and reduced their cholesterol levels". "Researchers suspect that this eating style works by increasing sex hormone-binding globulin, which binds to and inhibits the effect of estrogen, which is a primary culprit in menstrual pain and discomfort". A vegetarian diet may be useful for those suffering from rheumatoid arthritis. Research conducted at the University of Kuopio in Finland showed that people with rheumatoid arthritis who ate an uncooked vegan diet that was rich in lactobacilli experienced a relief of symptoms. After three months, when they returned to their meat-eating diets, the symptoms returned. In another study, researchers examined how people with rheumatoid arthritis responded to 7 to 10 days of fasting, which was followed by 3.5 months of a gluten-free, vegan diet, which was, in turn, followed by 9 months of a lacto-vegetarian diet. "For all clinical variables and most laboratory variables measured, the 27 patients in the fasting and vegetarian diet groups improved significantly compared with the 26 patients in the control group who followed their usual omnivorous diet throughout the study period".

While most cancer patients are told to eat whatever they wish, Keith Block, medical director of the Block Medical Centre in Evanston, Illinois, believes that this is a terrible mistake. He contends that fatty foods sanctioned by the National Cancer Institute support the cancer while starving the patient. In a cancer patient's body, "Metabolism speeds up as the immune system marshals its forces to fight the disease. Rapidly growing tumors become sinkholes for nutrients and calories. The body churns through protein, calories, vitamins and minerals, impairing immunity and leaving the patient increasingly susceptible to infection and unable to tolerate chemotherapy, radiation or drugs. Eventually, the cancerous tumors prevail, and severe malnutrition often leads to death".

Block contends that dietary recommendations should be part of cancer treatment, and he advises his patients to adhere to a strictly low-fat, vegetarian diet. More than half of all calories should come from complex carbohydrates; less than 15% of calories should come from fat. "Plant foods inhibit cancer. Phytochemicals in broccoli, cauliflower and other cruciferous vegetables help cancer-fighting enzymes purge carcinogens in cells. Garlic strengthens the immune system. Soyfoods contain genistein, a chemical that counteracts hormonal cancers and shrinks tumors. Carotenoids in orange colored fruits and vegetables as well as leafy greens shield cells from careening free-radicals that can cause cancer".

There is research supporting Block's beliefs. One study conducted in Sweden found that breast cancer was more likely to reoccur two to four years after surgery in patients whose diets were higher in fat. A review of men with prostate cancer determined that men with advanced prostate cancer ate more saturated fat than a control population, indicating that animal fat could promote the growth of malignant tumors.

Likewise, "Several epidemiological studies show a strong correlation between per capita fat intake and mortality from breast and other hormonal cancers, particularly for postmenopausal women. Other studies show that Italian and Asian women, who eat lower on the food chain and consume relatively little animal fat or protein, live longer with advanced breast cancer than American women". While chemotherapy and radiation have been known to take a devastating toll on patients, Block has found that his patients manage the treatments with fewer ill effects. He is convinced that the pHytochemicals in cruciferous vegetables detoxify the blood and flush poisonous residues of drugs and chemotherapy from the body.

Not everyone agrees that vegetarian diets are beneficial. Some people maintain that a vegetarian diet has limitations. A For 12 weeks, 9 of the men ate their usual meat-based diet. During the same time period, 10 men followed a lacto-ovo-vegetarian diet. The men who ate meat "experienced greater gains in fat-free mass and skeletal muscle in response to resistance training than those who ate a lacto-ovo-vegetarian (LOV) diet". Women who want to give birth to a son may need to think twice about vegetarianism.

"Researchers at Nottingham University in England have found that mothers who don't eat meat or fish are more likely than carnivores to give birth to daughters. In the general British population, boys outnumber girls at birth 106 to 100, but girls outnumbered boys 100 to 81 among babies born to vegetarian moms in the study". The prime concern appears to be that vegetarian diets may fail to supply sufficient amounts of certain vitamins. A study of elderly Chinese vegetarians, with a mean age of 81, found that the vegetarian diet appeared to be useful in preventing chronic illness. However, the diet was deficient in several B vitamins as well as "lower in total energy, fat and protein calories". Similarly, research published in the American Journal of Clinical Nutrition compared the nutritional intake and iron status of 50 vegetarian women between the ages of 18 and 45 to 24 age-matched meat-eating women. The vegetarians were found to have a relatively low intake of hemoglobin iron. Lacto-ovo-vegetarians "more frequently reported anemia during pregnancy, but this was not associated with any adverse pregnancy outcomes".

The USDA Grand Forks Human Nutrition Research Centre in North Dakota compared the zinc content in the diets of healthy young women who followed a lacto-ovo-vegetarian diet to women of a similar age who ate meat. "The lacto-ovo-vegetarian diet was associated with a 21% reduction in absorptive efficiency that, together with a 14% reduction in dietary zinc, reduced the amount of zinc absorbed by 35% (2.4 compared with 3.7 mg/d) and reduced plasma zinc by 5% within the normal range. Zinc balance was maintained with both diets." The researchers concluded that "although there is a greater risk of zinc deficiency in persons consuming lacto-ovo-vegetarian compared with omnivorous diets, with inclusion of whole grains and legumes zinc requirements can be met and zinc balanced maintained".

Whether vegetarians, especially vegans, consume adequate amounts of calcium is not clear. Some studies imply that the intake may be too low. On the contrary, many plants have high amounts of calcium. "One cup of turnip or collard greens or one-half cup of sea vegetables, such as hijiki, nori or kelp, contains about as much calcium as a cup of milk. In addition, calcium fortified rice and soy milks, tofu, dried fruits and sesame seeds are valuable non dairy sources of calcium".

The concern is not only about calcium intake, but also about how much is excreted. Some foods have the potential to increase the excretion of calcium. "For instance, it's believed that high-protein consumption interferes with calcium absorption, which means that dairy products can actually leach calcium even as they provide it. And dietary fibre will bind to calcium, effectively blocking its absorption-so it's best to eat high-fibre foods, such as enriched cereals, at separate meals from calcium-dense foods".

Moreover, to maintain calcium levels in the blood, the body requires levels of certain nutrients. "Vitamin D helps regulate calcium absorption from the digestive tract, slowing the excretion of calcium and channeling it back into the blood and bones. PhospHorous plays a delicate balancing act: When its levels are too high, calcium is excreted, a good reason to avoid high-pHospHorous products, such as soda. And other minerals, such as potassium and magnesium, are equally important in regulating calcium levels in blood and bones".

A number of studies have found that people who eat a plant-based diet are more at risk for vitamin B12 deficiency. The consequences of such a deficiency "include a host of neurological problems that range from subtle tingling of extremities, to actual loss of sensation or paralysis, to changes in memory. In fact, some of the symptoms commonly attributed to senility in old age are actually believed to be the result of B12 deficiency".

Is a vegetarian diet safe for children? Spock supported the diet for children, but others disagree. Concerns centre on the ability of small children to consume enough bulky plant foods to take in all the calories and nutrients they need to sustain normal growth. Particularly worrisome are the calcium and iron that are best absorbed from milk and meats, respectively. " Brody added that most experts recommend "a less radical approach to children's diets". Perhaps red meat and poultry may be eliminated, but fish, eggs, and diary products should be included.

How may problems be avoided? First, extra attention must be devoted to food planning. Without such care, "Food intake may be hapHazard, or lack variety, be higher in fat than is desirable, or be short on nutrients or calories." Also, it may be advisable to ask your medical provider to screen for iron-deficiency anemia. Vitamin C intake should be encouraged. "Foods that contain vitamin C, such as citrus fruits, improve the absorption of nonheme iron the type of iron in plant foods. Vegetarians should be advised also to limit intake of foods that are high in substances such as pHytates (whole grain, bran, and soy products) and oxalic acid (spinach). These substances decrease the absorption of iron".

If dairy products have been eliminated, calcium and vitamin D should be taken in supplemental form. Parents should also be certain that their children consume adequate amounts of vitamin B12 and zinc. The consequences of a B12 deficiency have already been noted. As for zinc, inadequate amounts may cause "delays in cognitive development." Further, "poor appetite and slowing of growth are the most evident clinical signs of zinc deficiency in children, particularly during infancy and adolescence".

The American Academy of Pediatrics maintains that some forms of vegetarian diets should be off limits for children. The Academy provides guidelines for infants, toddlers, and older children. Infants are potentially at greatest risk from inappropriate restrictions of diet. They are particularly vulnerable during weaning if they are fed a macrobiotic diet. Providing calorically dense foods at the time of wearing is important so that the increased bulk of vegetarian diets does not interfere with adequate consumption of energy, protein, and other nutrients. Toddlers are also unable to tolerate bulky diets. Meal plans must include foods of high caloric density, such as nuts, olives, dates, and avocados. Older children and adolescents on vegetarian diets are at less risk for growth failure than younger children. However, adolescents must understand basic principles of food selection. Without adequate nutrition knowledge, adolescents who combine vegetarian eating habits with weight-loss schemes are vulnerable to malnutrition and growth failure.

Adolescents who consider themselves vegetarians also have higher rates of eating disorders. In a study published in the August 1997 issue of the Archives of Pediatrics and Adolescent Medicine, researchers reported that "vegetarians were more likely to engage in all of the disordered eating behaviors examined in this study, which included frequent dieting, binge eating, self-induced vomiting and laxative use.... Vegetarianism may be serving as a marker for potentially harmful weight control behaviors....

It is possible that the adolescent is using vegetarianism as a socially acceptable way to avoid fat intake and to reduce energy intake". Ann Louise Gittleman, an award-winning nationally recognized nutritionist, does not think that a vegetarian diet is appropriate for everyone. "Some people, because of their genetic inheritance and ancestry, are better designed biologically than others for a vegetarian or near vegetarian diet. However, I believe everyone would be more healthy physically if they ate some meat. "

During her college years, Gittleman went on a vegetarian diet. Her response was far from ideal. "By about the middle of my senior year, my hair started falling out and I couldn't stand the noise in the dorm. My nerves were on edge, I lost about 20 pounds, and my skin broke out. My parents were very upset at my deteriorating pHysical condition, but no one could convince me to eat animal products. " Eventually, Gittleman visited a pHysician who specialized in nutrition. "He put me on megavitamins and insisted I begin eating chicken. That was the first step on my road back to health".

CONCERN FOR ANIMALS

People follow a vegetarian diet for a wide variety of reasons. For some, it is an integral part of their religious beliefs. The Seventh-Day Adventist Church and some Eastern religions, such as Buddhism and Hinduism, promote vegetarian diets. Some Jewish people believe that Judaism supports eliminating meat. Some of the greatest Jewish sages have taught that permission to eat meat was given later only as a grudging concession to people's weakness, and that many prohibitions and restrictions were applied to keep alive a sense of reverence for life". Further, in Judaism and Vegetarianism, Schwartz said that concern for animals is part of the Jewish belief systems. "The Jewish tradition clearly indicates that we are forbidden to be cruel to animals and that we are to treat them with compassion.

These concepts are summarized in the Hebrew pHrase tsa'ar ba'alei chayim, the biblical mandate not to cause pain to any living creature. Animals are also God's creatures, possessing sensitivity and the capacity for feeling pain; hence they must be protected and treated with compassion and justice". Apart from religion, there are many who are simply against killing animals for food or eating animal-based products. When there are so many alternatives, why should animals die-or be forced to live under terrible circumstances so that humans may eat? Often, these same people are opposed to various aspects of the meat-producing industry. Some believe that consuming animal products harms the environment.

Raised on a dairy farm in Montana, Howard F. Lyman was the fourth generation of his family to work as a farmer and cattle rancher. For 20 years, he also ran a feedlot. Lyman loved to eat meat. "Sure, I used to enjoy my steaks as much as the next guy," he wrote in Mad Cowboy. However, Lyman no longer eats any meat. In fact, he went on to become president of the International Vegetarian Union. He explains that the rendering industry uses the parts of slaughtered cows not eaten by humans, along with the entire bodies of diseased cows and other farm animals, euthanized pets, animals captured by animal-control agencies, and animals killed by motor vehicles.

The resulting mixture is ground and cooked. The fatty material floating to the top is refined for use in such products as cosmetics, lubricants, soaps, candles, and waxes. The protein material is dried and pulverized into a brown powder. About one-fourth of it consists of fecal material. This powder is used as an additive to pet food and livestock feed. In August 1997, the FDA issued a regulation banning the feeding of protein from ruminants (cud-chewing animals) to other ruminants. While cattle no longer eat solid parts of other cattle, sheep, or goats, they still are few ground-up parts of other animals and fowls, as well as blood and fecal material of their own species and of chickens. The environment in which animals live before they are killed may be far from ideal.

In Being a Vegan, Stefanie Iris Weiss, a writer and longtime vegetarian, writes that animals raised in agribusiness factory farms often live in a harsh, overcrowded environment that has devastating effects on animals. Chickens may try to peck each other to death. To prevent this, they may be debeaked. "Pigs in captivity also have been known to practice self-mutilation in the form of tail biting. Calves intended for slaughter are often kept in pens so small that their muscles are unable to develop properly.

This is seen as desirable because veal, the meat that comes from calves, will therefore be more tender." Moreover, Weiss contends that the meat industry wastes natural resources. "It takes 2,500 gallons of water to produce one pound of meat-both to feed the animal and grow the food that the animal consumes. Water is not an endless resource, and much of the ground water left on the earth is contaminated. Factory farming also has contributed to water contamination, for a great deal of animal excrement and pesticide residue has been released into rivers. Cattle raised for meat production consume enough grain to feed 8. 7 billion people a year".

In Prisoned Chickens, Poisoned Eggs, Karen Davis describes standing outside a chicken slaughter plant and watching the trucks with thousands of chickens, who may wait in the trucks for several hours before they are processed. "The killing of birds normally involves three pHases: motor paralysis by means of an electric current (inaccurately called "stunning"), throat-cutting, and bleeding. Poultry slaughtered are neither stunned (rendered unconscious) nor anesthetized (rendered pain-free). Pre-slaughter stunning of poultry is not required by law and is not practiced despite the use of the term "stun" to denote what is really immobilization by means of painful electric shocks".

Davis also details the chickens' pain and suffering: "Like mammals subjected to painful stimuli, chickens show a rapid increase in heart rate and blood pressure, and behavioral changes consistent with those found in mammals indicating pain perception-efforts to escape, distress cries, guarding behaviour, and the passive immobility that develops in birds and other animals subjected to traumatic events that are aversive and that continue regardless of attempts by the victim to reduce or eliminate them". An organization dedicated to reducing animal cruelty, outlines other negative aspects of the poultry-farming industry. For example, it notes that almost all the laying hens in the United States live in battery cages, which are "lined up in rows and stacked in tiers inside of huge warehouses." Poultry.org, states that poultry that are bred for eating also have dreadful lives. Most chickens and turkeys are crowded into sheds called "grower houses. " Chicken and turkeys bred for meat have been genetically altered to grow faster and larger. This rapid growth rate increases profitability, but also causes health problems, such as legs that cannot support the birds' bodies.

"Birds raised in the United States for meat-mainly chickens and turkeys-may be sold as 'range' if they have U.S. Department of Agriculture certified access to the outdoors. No other criteria, such as environmental quality, size of area, number of birds, or space per bird are included in this term. To date, there is no legal or commercial definition of husbandry terms regulating the sale of eggs in the U.S. There are no standards governing the term or the claim 'range' or similar advertisements on egg cartoons, such as 'free running, ' 'free roaming,' or 'free walking.'"

Moreover, "free-range hens are debeaked at the hatchery the same as battery-caged hens. Debeaking is a painful mutilation that impairs the hens' ability to eat normally and to preen (practice bodily hygiene). Typically, 2,000 or more hens-each hen having only one to two square feet of floor space-are confined in a shed without access to the outdoors during their lives. If the hens can go outside, the exit usually is very small allowing only the closet hens to get out The yard may be nothing but a mud yard saturated with droppings and intestinal coccidia and other parasites".

Veal calves are slaughtered in lots, each lot containing ten to twenty veals. The calf is three to four months old at slaughter and weighs between 250 and 300 pounds. Veals are kept off feed for eighteen hours before slaughter and handled with care to avoid bruising. Veals have to be dragged to the slaughterhouse, as they are too weak to stand and walk.

The animals are forced into the restraining pen, two at a time. The veals can see their comrades having their throats cut. The calves' eyes become practically white. Foam is pouring out of their mouths. The man with the stun gun waits until there is enough space in the conveyor belt to receive the newly stunned veals. A large metal bolt strikes through the animal's skull.... The veal is hung upside down by a chain, and the throat is slit. The animal bleeds to death, which takes up to five minutes, as it moves along the conveyor belt. The head is taken off. The body is slit open and steam comes out with all the entrails. Hooves come off. The carcass is cut into parts. Here is what Coe observed on slaughter day at an egg factory in North Carolina: Two men go down the centre aisle, grabbing hens by legs, wings, and necks, and stuffing them in crates. The hens are in total dread and terror. They make terrible screeching sounds. The crates are then thrown on top of each other and are left to wait in the cold for the truck.

By the time the hens get to the slaughterhouse, eighty percent of their bones have been broken. The bones are very brittle, as all calcium goes to the eggs.

For the vast majority of pigs, the reality is far different. In Vegan, Erik Marcus, a writer and public speaker, noted that in the 1980s big corporations assumed control over the pig industry and applied the same principles that are in place for poultry farming. For the pigs, the pHysical living conditions have markedly deteriorated. "Pig sheds are designed to hold the largest number of animals at the lowest cost. Concrete slats are the standard flooring in commercial pig operations. Sleeping on concrete is more than uncomfortable-over time, it creates serious health problems. Joints swell, skin gets scraped off, and the feet get serious abrasions and infections. This adds to the pigs' stress and increases rates of fighting and cannibalism. " Many pigs become ill, especially from pneumonia, which may be caused by air that is heavy with dust and smells of ammonia.

What is life like for a typical pig in a commercial farm? "Shortly after birth, workers snip notches out of each piglet's ear for identification purposes. No anesthetic is used. To reduce injuries caused by fighting, their 'needle' teeth are clipped again without anesthetic.... Pigs are raised in as little space as possible. For young 250-pound pigs, operators are advised to allow a little more than one square yard of floor space for each animal. The crowding is not just to save space-crowding also reduces feed costs. A modern pig farmer wants his pigs to stay as motionless as possible-when a pig walks, the farmer sees costly feed wasted to provide the energy for movement instead of being stored as flesh". Pigs often panic during the rush to get them loaded on the truck that will transport them to the slaughterhouse.

Overloaded trailers cause extra heat, fighting, and stress that kill many pigs. Disputing many of these claims, the American Meat Institute Foundation notes that legal requirement and ethical and economic considerations obligate meat packers to handle livestock in the most humane manner possible. The Humane Slaughter Act of 1978 established strict standards for packing plants that are monitored by federal meat inspectors nationwide. A violation of the act can result in almost immediate shutdown of a plant. The Humane Slaughter Act requires that

- Animals be handled and moved through chutes and pens in ways that do not cause stress;
- Livestock must be rendered insensible to pain prior to slaughter;
- Animals must have access to water and those kept longer than 24 hours must have access to feed;
- Animals kept in pens overnight must be permitted room to lie down;
- Downers or crippled livestock must not be dragged in the stockyards, crowd pen, or stunning chute.

The American Meat Institute Foundation states that the meat industry has taken a number of initiatives to ensure that animals are handled properly. In 1991, it published Voluntary Animal Handling Guideline for Meat Packers, which was translated into Spanish in 1992. In 1997, the industry published Good Management

Practices for Animal Handling and Stunning. In 1999, the American Meat Institute Foundation held its firstever Animal Handling and Stunning Conference and released two training tapes: "Good Animal Handling Practices for Beef Processors" and "Good Animal Handling Practices for Pork Processors".

To the American Meat Institute Foundation, "humane slaughter is an ethical imperative and a common-sense business practice." Plant workers do all that they can to minimize the amount of discomfort animals experience during the slaughter process. "When the slaughter process is performed correctly, the process is so rapid that livestock never feel pain." Further, humane healing makes economic sense. When animals are stressed or treated poorly, the meat is of lesser quality.

CHAPTER 6

Organic Foods

Advertising tries to persuade consumers that organic food is healthier and worth the extra price. Whether organic foods indeed provide the benefits claimed for them is a matter of controversy. In the supermarket, there is a large display of apples that sell for 99 cents per pound. There is also a similar arrangement of organic apples. However, these retail for $1. 49 per pound. Are the organic apples worth the extra 50 cents per pound? Are they better than the regular apples? Until the 1970s, there were few commercially available organic foods, most found only at small co-ops.

That has all changed. Now organic foods are available at regular suburban grocery stores as well as specialty shops and are purchased by consumers from all ethnic backgrounds and economic classes.

Organic foods are at least a six-billion-dollar business that keeps growing. "Sales of organic food in North America have been increasing 20 to 30 percent annually over the past decade, and organic food is now the fastest-growing segment of the natural-foods market". Who is buying organic foods? Read what the Organic Trade Association as to say:

- A study conducted in March 2001 found that, at some point, 63% of Americans buy organic foods and beverages.
- Organic users tend to be 31% more concerned about the environment and pollution than the general population.
- More than 40% of the people who use organic food are between the ages of 36 and 55. People who use organic products are 25% more likely to have a bachelor's or graduate degree.

STANDARDS FOR ORGANIC FOODS

What makes a food organic has little to do with the food; instead, it is primarily a function of how the soil was prepared and how the food was produced. "In fact, to understand what you're getting when you buy organic, you have to forget fruits and vegetables for a moment and think about dirt". The techniques organic farmers use to control weeds and insects have existed since the beginning of agriculture: crop rotation, cultivation, mulching, soil enrichment, and the encouragement of natural predators and microorganisms that keep pests in check. The report added that farmers who produce organic products "are allowed to use only natural pesticides: insecticidal compounds extracted from plants, bacteria that attack crop pests, soaps, and a few types of metals and minerals.

In order to be considered organic, "crops must be grown without chemical or biological interference, without genetic modifications, on land that is free of contaminants. For animal products, it means animals raised for food must be treated in a humane manner, not confined, not genetically altered, and not given hormones or antibiotics that might show up in meat or dairy products". This type of farming increases labour costs and makes organic foods more expensive.

The soil teems with microorganisms that give vigour and resistance to plants, with worms to aerate the soil and insect predators to control pests. High in humus and nurtured with natural fertilizers, a living soil provides plants with strength to resist disease and insects, as well as superior taste and nutritional value". Organic growers manage their farms as whole systems. They focus on improving soil health and use farming methods that prevent problems, such as rotating crops from field to field to manage pests and weeds and to improve soil health and fertility, using animal manures and vegetative matter as fertilizers, monitoring fields regularly to determine when pest or weed control should be done, and recycling and composting organic wastes.

For many years, certification of organic farmers was conducted by more than 40 state and private organizations. The new standard, which required more than 10 years to develop and did not become fully effective until mid-2002, prohibited "the use of irradiation, biotechnology and sewer sludge fertilizer for any food labeled organic." It also forbade "synthetic pesticides and fertilizers in the growing of organic food, and antibiotics in meat labeled organic".

Fig. Organic Food

Further, dairy cattle must "have access to pasture". Foods that meet the standard's criteria are labeled "USDA Organic." A USDA news release noted that "all agricultural products labeled organic must originate from farms or handling operations certified by a state or private agency accredited by the USDA. Farms and handling operations that sell less than $5,000 worth per year of organic agricultural products are exempt from certification".

The standard also established four categories for labeling. Foods that are labeled "100 percent organic" must contain only organic ingredients. Ninety-five percent of the ingredients of any food labeled "organic" must be organic. Foods that contain at least 70% organic ingredients may contain a label that says, "Made with Organic Ingredients." Three of the ingredients may be noted on the front of the package. Finally, "Processed products with less than 70 percent organic ingredients may list organic ingredients on the information panel but may not carry the term 'organic' anywhere on the front of the package".

When the standard was announced, Dan Glickman, then the secretary of Agriculture, noted, "Let me be clear about one thing. The organic label is a marketing tool. It is not a statement about food safety. Nor is 'organic' a value judgment about nutrition or quality. USDA is not in the business of choosing sides, of stating preferences for one kind of food, one set of ingredients or one means of production over any other. As long as rigorous government safety standards are being met, we stand ready to do what we can to help support any farmer and help market any kind of food".

More than a decade ago, the Organic Foods Protection Act (OFPA) of 1990 was passed. It mandated that the USDA create national standards for organically produced agricultural products. To be sold as organic, products needed to adhere to these standards. "The OFPA and the National Organic Programme (NOP) require that agricultural products labeled organic originate from farms or handling operations certified by a State or private agency that has been accredited by the U.S. Department of Agriculture (USDA). The OFPA and proposed regulations do not address food safety or nutrition. The USDA Agricultural Marketing Service, the part of the USDA that set marketing standards, is charged with implementing the NOP".

In December 1997, Glickman "proposed federal organic standards for U.S. food crops that included... genetic engineering, the use of sewage sludge as fertilizer, irradiation and livestock confinement practices". It is believed that the proposed standards were created following much lobbying from the giant businesses. How else, for example, could one explain the addition of sludge? "Sewage sludge contains some 60,000 toxic chemicals, a mix of residential, industrial and commercial discharge that includes hospital wastes, street runoff, heavy metals, PCBs, dioxin, solvents, asbestos and radioactive wastes".

New standards that prohibited the use of genetic engineering and addressed the other contested issues were proposed in March 2000. All products labeled "USDA Certified Organic" must "be produced without toxic pesticides or toxic 'inert ingredients;...antibiotics, growth hormones, and rendered animal protein cannot be administered or fed to animals;... intensive confinement of livestock will not be allowed; and... no synthetics or chemicals will be allowed in organic production without the approval of the NOSB [National Organic Standards Board]".

PESTICIDES

Organic farmers use no synthetic pesticides. Pesticides are poisonous compounds that are designed to kill elements of nature that damage crops, including insects and fungal pests. As such, they have the potential to cause a variety of medical problems. Information about the effects of pesticides has primarily been obtained from studies of pesticide applicators and farmworkers. Common reactions to pesticides include nausea, lung and eye irritation, and temporary nerve damage. Long-term and chronic medical problems have been harder to verify, but farmers using more pesticides seem to be at greater risk than nonfafmers for some forms of cancer and amyotropic lateral sclerosis (Lou Gehrig's disease), and tests on animals indicate that some pesticides may cause birth and immune-system defects.

It is suspected that many pesticides may interfere with hormone systems that regulate functions like reproduction. "For consumers in general, the unsettling truth is that no one really knows what a lifetime of consuming the tiny quantities of pesticides found on foods might do to a person. The effect, if any, is likely to be small for most individuals-but may be significant for the population at large".

In The Organic Gourmet: Feast of Fields, Tracy Kett, a writer, caterer, and promoter of organic agriculture, noted, "Several studies have found that the possible short- and longterm human health effects of agricultural chemicals are numerous and range from respiratory problems in field workers (despite hazardous working conditions, farmers and their employees often work without any kind of safety clothing or respiratory protection) to severe allergic and asthmatic reactions, reproductive disorders, cancer and degenerative diseases in both consumers and farm workers. " Most of these chemicals have no purpose.

In Eating with the Seasons, Paula Bartimeus, a professional nutritionist and regular writer for health magazines, said that "although the Government states that the levels of pesticides used are well within the safety margins, random testing has proved otherwise. Many farmers mix chemicals before spraying, and the combined and cumulative effects of such concoctions are still uncertain".

The author of Designer Poisons, Marion Moses, lists some precautionary statements from pesticide labels: "Avoid inhalation or contact with eyes or skin. Harmful if swallowed. May cause eye irritation." "Avoid eye contact. Wash thoroughly after handling." "Corrosive to eyes. Causes eye damage. Do not get in eyes. Harmful if swallowed. Avoid contact with skin or clothing." "Do not apply to excessively sunburned or damaged skin. May cause skin reaction in rare cases." "Do not get in eyes, on skin, or clothing. Harmful if swallowed, inhaled or absorbed through skin".

The Environmental Protection Agency (EPA) regulates pesticides, but there is concern that the regulation is inadequate. A number of groups concerned about the use of pesticides believe that the EPA "has not tested all the ingredients in these chemicals and does not require companies to disclose or label their so-called inert ingredients." Moreover, the EPA fails to "take into consideration the effects of combinations of pesticides (only the effects of individual pesticides have been studied)." Further, there is no acknowledgment that almost all pesticides "drift from their point of application".

If a pesticide, applied according to the package label, doesn't result in residues higher than a safe level, that safe level becomes the legal limit. But the legal limit can sometimes exceed the level that safety alone would require if the government decides the benefit to farmers outweighs the risk to consumers". Another concern is that "because government standards for pesticide levels are based on the height and weight of an adult man, it's unclear whether pesticide consumption at those levels might be harmful to a child".

A five-year study completed by the National Academy of Sciences found that children were more vulnerable to pesticides than adults. Nutritionist Susan M. Kleiner offered several reasons. "Children eat more food in proportion to their body weight than do adults, so their exposure to pesticides in greater. Children eat more fruit... compared with adults. Children will often eat a lot of one or two specific foods. If that food has pesticide residues, it will increase their exposure to pesticides".

The Environmental Working Group (EWG), a leading content provider for public interest groups with a special interest in the effect pesticides have on children, arranged for laboratory evaluations of baby foods produced by Gerber, Heinz, and Beech-Nut. The testers "found 16 different pesticides in eight baby food products tested. Of those 16 pesticides, three are probably human carcinogens, five are possible human carcinogens, eight are neurotoxins, five are endocrine disruptors, and five are categorized as oral toxicity 1 chemical, the most toxic designation. More than half of all samples (53 percent) contained detectable levels of pesticides, 18 percent of samples had two or more pesticides in them, and one sample contained three different pesticides".

The EWG obtained data from the U.S. Food and Drug Administration on the pesticide content of 42 fruits and vegetables. "More than half of the total dietary risk from pesticides in these foods was concentrated in just 12 crops. The pesticides that were found in these foods are classified by the Environmental Protection Agency (EPA) as probable human carcinogens, nervous system poisons and endocrine system disrupters." The 12 most contaminated foods, in order of levels of contamination, were strawberries, bell peppers (green and red) and spinach (tied), cherries (U.S.), peaches, antaloupe (Mexican), celery, apples, apricots, green beans, grapes (Chilean), and cucumbers.

It should not be surprising that there has been a boom in the sale of organic baby food. At least one school system- Berkeley, California is offering pesticide-, herbicide-, and hormone-free foods to its students.

"For now, schools are buying their goods from Berkeley's twice-a-week farmers' market, from where organic farmers and vendors are regularly invited to visit with students to explain their natural farming techniques. Educators hope to integrate gardening know-how with classroom education, and, in the future, individual schools plan to start up organic gardens of their own".

One cannot just wash off the pesticides before preparing the food, because some pesticides are not water soluble. Also, "Many chemicals routinely used in farming can penetrate the fruit or vegetable skin and therefore can't be washed off. Peeling helps remove them but eliminates valuable nutrients in the process. And... a wax coating... not only seals in pesticides, but the wax itself may contain fungicides, coloring agents and other potentially unhealthy substances".

LIMITATIONS OF ORGANIC FOODS

Organic foods, however, are not necessarily healthful or safe. A high-fat food that is made from organic products remains high in fat. "You can buy organic chocolate bars, ice cream and cookies-all made with ingredients that are pesticide-, chemical-, antibiotic- and hormone-free-but they'll be laden with fat, sugar and calories". Moreover, the manure that organic farmers may use could contain bacteria. "The only real difference between organic and nonorganic food is in the growing-and that's not a big enough difference to protect your health from bacteria."

Like other forms of food production, there may be compromises in the healthfulness of the organic food environment. In a January 1999 issue of the Humanist, Lisa Hamilton wrote that the major producers "may dump obscene amounts of quick-to-leach chicken manure on poor soil, carrying hazardous nitrates directly to groundwater, or blast crops with the nonexclusive, plant-based pesticide Pyrethrum, killing most insects in range, including beneficial ladybugs and lacewings". Nevertheless, a May 2000 survey of 1,029 men and women commissioned by the National Centre for Public Policy Research, a Washington, DC, nonprofit education foundation, found that the majority of the public believe that products labeled "USDA Certified Organic" would be "safer," "better," and "healthier for consumers" than non organic foods.

Responding to the results of this study, John Carlisle, director of the Environmental Policy Task Force at the National Centre for Public Policy Research, said that misperceptions about the benefits and risks of organically certified products would be heightened by this USDA proposed label. "Clearly, consumers want the USDA to amend this rule to include specific language on the USDA proposed seal to inform consumers that organic certification is based on production methods and conveys no assurance of food safety, nutrition or other quality".

As a result of this study, the National Centre for Public Policy Research advised against the creation of a USDA organic seal that would only impart misinformation and create confusion. A letter from John E. Frydenlund, the director of the taxpayer watchdog group Citizens against Government Waste, to the programme manager of the federal government's National Organic Programme stated that "there is no scientific evidence that organic food is safer, healthier, or, in any way, better than non-organic food." Moreover, Frydenlund continued, there are "inadequate safeguards to protect consumers from the health dangers caused by composted and raw manure used in organic production. This inadequacy is exacerbated by a failure to provide adequate information informing consumers of the particular health dangers that exist in the consumption of organic foods".

There are some additional limitations to organic farming. "Because organic farmers eschew pesticides and herbicides used by conventional farmers, they cannot use modern conservation tillage techniques that have been extraordinarily successful in reducing soil erosion. As a result, compared to conventional farming, organic farming is woefully inadequate in controlling soil erosion.... But the most glaring environmental disadvantage of organic farming is its exorbitant need for land. Organic farming is only about half as productive as conventional farming, which means organic farming requires far more land to feed the world than modern methods". In an interview published in the April 2000 issue of Reason, Nobel Peace Prize winner Norman Borlaug, an agronomist, noted, "Even if you could use all the organic material that you have the animal manures, the human waste, the plant residues and get them back on the soil, you couldn't feed more than four billion people.

In addition, if all agriculture were organic, you would have to increase cropland area dramatically, spreading out into marginal areas and cutting down millions of acres of forests".

These are essentially the beliefs of Dennis T. Avery, an expert in international agriculture, in Saving the Planet with Pesticides and Plastic. "With present knowledge levels, no responsible authority or organization should recommend either organic farming or traditional low-yield farming systems as a broad-gauge alternative to high-yield agriculture. In fact, slashing farm chemical usage is likely to produce more soil erosion, more human cancer, and less wildlife habitat. At present, organic farming could not even sustain the fertility of our existing cropland, or protect it effectively from erosion".

CHAPTER 7

Genetically Modified Foods

Increasingly, genetically modified foods are making their way to the marketplace. Since they are not labeled, consumers generally do not know that they are eating such products. Not everyone is comfortable with the present system.

A few years ago, one rarely heard about genetically modified organisms (GMOs). In fact, few people knew exactly what they were. That is no longer the case. GMOs, or plants, animals, or microorganisms that have been genetically altered or engineered for a specific purpose, are everywhere. It has been estimated that almost half of all American farmers grow some variety of genetically modified crops, also known as genetically engineered crops.

The Food and Drug Administration (FDA) says that GMOs are filling the shelves of stores. "Tomatoes, potatoes, squash, corn, and soybeans have been genetically altered through the emerging science of biotechnology. So have ingredients in everything from ketchup and cola to hamburger buns and cake mixes".

In Genetically Engineered Food, Martin Teitel, executive director of the Council for Responsible Genetics, and Kimberly A. Wilson, director of the council's programme on Commercial Biotechnology and the Environment, describe the sweeping changes brought about by GMOs. "The genetic engineering of our food is the most radical transformation in our diet since the invention of agriculture 10,000 years ago. These food crops are already growing on millions of acres all around the world".

Similar sentiments were expressed in Genetically Engineered Food by Ronnie Cummins, national director of the Organic Consumers Association, and Ben Lilliston, a health and environment writer and communications coordinator for the Institute for Agriculture and Trade Policy in Minneapolis: "Genetic engineering... is a revolutionary new technology still in the early experimental stages of development. It enables molecular biologists to permanently alter the essential characteristics or genetic codes of living organisms. This technology has the awesome power to break down fundamental genetic barriers-not only between species but between humans, animals, and plants.

For the first time in history, the scientists and corporations using this technology have become, in effect, the architects, builders, and 'owners' of life". Introducing the genes of one type of food into another, thereby changing the genetic structure of an organism, gives the second plant or animal beneficial characteristics, such as resistance to disease, improved nutritional value, and better growth. These modifications enable foods to grow faster, stronger, or bigger, all while using far less pesticide. There are now tomatoes that last longer and corn that is resistant to pests. "The acreage devoted to herbicide-resistant crops has been expanding because planting them reduces the need to plow more ground, decreases the amount of herbicidal chemicals needed, produces higher yields, and can deliver a higher grade of grain and other products".

Transgenic or genetically modified crops "can be developed for pest resistance, improved yield, tolerance to biotic and abiotic stresses, use of marginalized land, improved nutritional content, and the production of vaccines and pHarmaceuticals". A 2000 story in Popular Science said, "Genes from flounders can help ordinary plants like tomatoes and strawberries fight the cold. Researchers are also inserting bacterial genes into corn and soybean plants to better protect them from insects or render them immune to certain herbicides".

Cross-breeding has been around for a long time. "Virtually all common fruits and vegetables look and taste the way they do because of hybridization". However, biotechnology has added a whole new dimension to selective breeding. "This process allows for the transfer of only one or a few desirable genes, thereby permitting scientists to develop crops with specific beneficial traits and those without undesirable traits. Current technology permits scientists to alter one plant characteristic at a time, thereby not spending years trying to develop the best tasting and hardiest plants".

For example, Craig Nessler, professor and head of the Department of Plant Pathology, Physiology, and Weed Science at Virginia Tech University, inserted genes from rats into lettuce seeds. This produced leafy greens with much higher amounts of vitamin C. "Rats, it turns out, carry a gene that allows their bodies to manufacture vitamin C, which is why the little four-legged fiends never develop scurvy on long sea voyages". Years ago, such science would have been unthinkable.

The differences between traditional breeding and genetic engineering. "To give a simple example, a traditional breeder interested in producing a yellow tomato must find the yellow trait in a plant that will breed with the tomato by natural mechanisms. The only plants that can breed with tomatoes are closely related ones. Unrelated plants like oak trees or cantaloupes could not breed with tomatoes, and thus could not contribute new genes. A genetic engineer, on the other hand, can consider any organism-even a butterfly or a daffodil as a source of the yellow trait. If the gene that determines yellow colour has been identified and isolated, it can be directly transferred into tomato plants".

The International Food Information Council maintains that because GMOs reduce the amounts of pesticide used by farmers, they are healthier for the environment. "Insect protected crops allow for less potential exposure of farmers and groundwater to chemical residues, while providing farmers with season-long control. Also by reducing the need for pest control, time, effort and resources spent on the land are less, thereby preserving the topsoil".

It is almost impossible for the average consumer to know whether there are GMOs in his or her food. Playing on consumers' fear of the unknown, a few innovative marketers include "No GMOs" notations on their labels. However, the vast majority of products do not indicate whether they are comprised of any genetically altered ingredients.

"Legal authority for food labeling rests with the Food and Drug Administration. Foods derived from biotechnology currently must be labeled only if they differ significantly from their conventional counterparts-for example, if their nutritional content or potential to cause allergic reactions is altered".

SUPPORTERS AND OPPONENTS OF GMOS

Should consumers feel comfortable eating genetically altered food? The controversy surrounding GMOs, or Frankenfoods, as they are called by those who oppose them, is filled with heated discussion. In general, the federal government, big business, and many researchers and numerous well-respected publications maintain that genetically altered food is perfectly safe. They contend that GMOs offer the potential to feed healthier food to far larger numbers of people. That is of particular significance, they say, when considering the vast amounts of food needed to feed the populations of developing countries. About 15% of the world's population-about 800 million people-consume less than 2,000 calories per day. They are always hungry and live in a state of chronic malnourishment.

Large numbers of these people are women and children. "More than 180 million children under five years of age are underweight, that is, they are more than two standard deviations below the standard weight for their age. This represents one-third of the under-fives in the developing countries." These groups are growing at a staggering rate. "By the year 2020, there will be about an extra 1.5 billion mouths to feed. If the proportion of the population of the developing countries deprived of an adequate diet remains the same, the number undernourished 20 years from now could be well over one billion".

Not everyone agrees. Sizeable numbers of equally esteemed researchers and consumer and environmental groups worry about a variety of factors. "Concerns include ethical issues related to potential long-term health effects of eating bioengineered foods, labeling, and potential environmental risks". In Europe, particularly Great Britain, large numbers of people have supported a ban of GMOs. Grocery stores are refusing to carry foods containing GMOs. Other countries are following Britain's lead.

Moreover, concern about risk of lawsuits and decline in land value has made rural land managers hesitant to authorize tenants to grow GM crops." Within the European Community, there is enormous apprehension. "Consumers worry about the impact of any health risks that might eventually be discovered, especially if whole national populations have ingested GM food. Health effects, moreover, may stem from risk related to questions that scientists have not yet asked rather than those questions for which reassuring answers have been found". In High-Tech Harvest, Elizabeth L. Marshall, a science writer, notes that the most serious GMO controversy has centered on food labeling. Usually, opponents of GMOs favour labeling. They want to know which foods have been changed and believe that shoppers, when presented with such information on the label, would shy away from them. "But some supporters want labels too. They believe that labels would help pave the way to consumer acceptance".

GMO HISTORY

Genetic engineering food research began in the 1980s. In 1990, the enzyme chymosin was the first genetically engineered food to receive the approval of the FDA. More than half of the hard cheese currently sold in the United States is made with chymosin produced by genetically engineered fungi. The rest is made with chymosin obtained from rennet, an enzyme from the stomachs of slaughtered calves. "Although cheese-makers can use chymosin from either source, the chymosin made from genetically engineered bacteria is easier to obtain and more pure".

Two years later, in 1992, the FDA approved the first genetically engineered whole food-the Calgene Flavr Savr tomato, which became available for sale in 1994. Because of a change to a single gene, the Calgene Flavr Savr tomato ripened without getting soft. However, the Flavr Savr tomato encountered a number of problems, such as an inability to tolerate shipping. "Contrary to Calgene's expectations, the tomatoes were often so soft and bruised that they could not be sold as fresh produce, and most of the... varieties did not have acceptable yields or disease resistance in tomato-growing regions". The Flavr Savr marketing campaign was unsuccessful.

The tomatoes received a good deal of negative press, and consumers refused to buy them. "Calgene even tried marketing the tomatoes as a gourmet product under the friendly sounding 'MacGregor's' brand name, but despite their presence in thousands of U.S. grocery stores, consumers did not want to pay more for genetically engineered tomatoes." In 1996, the Flavr Savr tomato was removed from the market. Teitel and Wilson noted that "the Flavr Savr grew well in the laboratory but encountered serious obstacles in the field". Nevertheless, genetic engineering moved forward. By 1996, genetically engineered seeds for corn and soybeans were available, followed by seeds for potatoes, squash, tomatoes, and rapeseed. "Each type of seed offers a characteristic that natural seeds did not. Some protect the plants against insects or disease, while others delay ripening or protect the plants from chemicals used to kill weeds.

REGULATION OF GMOS

It is important to realise that genetic engineering is a tool that may be used in a variety of ways to make many different food products. That is why, Marshall contends, the FDA decided in 1992 that genetically engineered foods would be reviewed "based on their individual safety and nutrition, rather than on the methods used to produce them." The FDA deemed them "as safe as foods developed through other agricultural technologies".

As a result, the FDA regulates GMOs as it does any other food. If a new product contains food items that are already considered safe, the manufacturer does not need to obtain special permission to sell the item. "In other words, a transgenic [another word for genetically engineered] catfish containing a trout gene would be considered safe to eat because it is already well-known that trout is harmless to humans.... The only time the law requires testing of a new food is when it contains an additive that is not known and, therefore, not generally regarded as safe".

In 1992, the FDA determined that there would be exceptions to this rule. If a new food contained ingredients that were likely to trigger allergic reactions in allergic individuals or if the manufacturing process dramatically altered the nutritional content, a label would be required.

Unfortunately, this is not as clear-cut as it may initially appear to be. Foods are constantly combined together to form new foods. Should everything be labeled? The government and sizeable numbers of people within the food industry say that such labeling would be confusing and expensive.

In Pandora's Picnic Basket, Alan McHughen, a senior research scientist at the University of Saskatchewan, Canada, wrote that "in North America, regulators and companies agree that mandatory labels should be reserved for those products carrying a documented health risk or substantial change in nutritional composition.

If the GM products are 'substantially equivalent' to conventional counterparts, the companies argue, the GM label would be misleading". Nonetheless, many consumer groups consider labeling to be an absolute necessity. EarthSave International explains that genetic engineering "introduces new proteins into the human and animal food chains. This means that human beings are now consuming products that have never before been considered foodstuffs.

There is concern that these new proteins could potentially cause toxic or allergic reactions or other health effects. Unfortunately, there is no easy way to predict the allergenic potential... since allergic reactions typically occur only after the individual consuming the food is sensitized by initial exposure to the allergen". It is true that the government has mandated the labeling of other products such as foods that are processed and those containing sulfites. Labels must note the source of hydrolyzed proteins, and they are required to state if a food-other than a spice-has been irradiated. Cigarette labels have a warning from the surgeon general. "These legal requirements are in place because large numbers of citizens want such information, and a specific fraction needs it. An identifiable fraction of consumers actually needs information about genetic modifications-for example, regarding allergenicity... and almost all consumers want it".

Lawrence Kushi, an epidemiologist at Columbia University, notes that it is simply untrue to say that genetically modified food is no different from the conventional equivalents. He maintains that labeling such foods would give consumers the information they want while allowing scientists to examine whether they have any negative impacts on human health or the environment.

ARGUMENTS IN FAVOUR OF GMOS.

Some people contend that the negative publicity that consumer groups generate about GMOs is most harmful to the world's most vulnerable people-the poor. They say that unlike residents of wealthier countries, large segments of the population of developing countries worry about dying from malnutrition and starvation. By using GMOs, farmers in developing countries may harvest more crops with less pesticide. Playing on this theme, frequently appearing television and magazine ads have featured rice that has been genetically altered with a daffodil gene. As a result of that modification, the rice contains beta-carotene, which the body converts into vitamin A. The ads claim that by eating this form of "Golden Rice," countless numbers of people in poorer countries will be spared blindness.

About 400 million people may be in danger of vitamin A deficiency. Of these, 100 to 200 million are children. Mammals do not manufacture vitamin A, so it must be obtained through the diet. People who eat lots of fruits and vegetables and animal products obtain adequate amounts. Millions of people are not as fortunate. "1.0 to 2.5 million deaths per year of preschool children-up to 30% of total deaths in that age group could potentially be averted by bringing vitamin A deficiency under control worldwide.... Combined with expanded vitamin A supplementation programs-which will continue to be important-Golden Rice is expected to make *a* major contribution to improving the health of millions of the world's poorest children".

For the unspoken challenge here is that if we don't get over our queasiness about eating genetically modified food, kids in the third world will go blind. " But Pollan asserts that there is far more to the Golden Rice issue than is initially evident. For example, in order to meet the minimum daily requirement for vitamin A, it would be necessary for an average 11-year-old to eat 15 pounds of cooked Golden Rice.

Moreover, in order to convert beta-carotene into vitamin A, fat and protein must be present, and these are lacking in the diet of malnourished children. Pollan also wonders whether Asians will accept rice that is golden in colour. It is already well known that brown rice is healthier than white rice, yet Asians consistently prefer white rice. Apparently, at some point, Confucius "extolled the pure whiteness of rice as the ideal backdrop for green vegetables". Pollan further questions the ethics of an industry that is using the suffering of children to sell its food.

In "Vitamin A Deficiency Disorders: Origins of the Problem and Approaches to Its Control, " Alfred Sommer writes that there are some "hurdles" to be overcome before Golden Rice may have an impact on society. "The strains must be able to grow under the varied conditions in countries with vitamin A deficient populations. The yield and the cost must be attractive to the farmer (or benefit from public sector subsidization). The organoleptic [sensory] qualities of the rice must be acceptable to the target population (women and children). The beta-carotene needs to be bio-available, the degree dependent on its concentration in the rice, the matrices to which it is bound, the effect of traditional cooking methods and the amount consumed". Sommer warns that Golden Rice will never be a complete solution to the problem of vitamin A deficiency. "Many deficient populations do not consume rice, and even within traditional rice-consuming countries, some high-risk groups will not be able to afford it".

Regardless, it is impossible to deny that the population of the world is. growing at a sometimes mind-boggling rate. Presently, the world has about 5 billion people. By the year 2050, that figure will double to 10 billion. Large numbers of these people will live in the developing world, where hunger is rampant. The International Food Information Council states that "by increasing a crop's ability to withstand environmental factors, growers will be able to farm in parts of the world currently unsuitable for crop production. Along with additional food, this could provide the economies of developing nations with much-needed jobs and greater productivity".

Feeding so many people takes a staggering toll on the environment. "Erosion can claim precious topsoil, farm chemicals sometimes reach streams, rivers and ground water supplies, and livestock can deplete grazing lands. Wetlands and other sensitive habitats sometimes get plowed under for use as farmlands. And, in the world's tropical forests where an estimated 90 percent of the world's species exist, poor farmers clear trees in order to provide food and a living for their families." The International Food Information Council believes that GMOs may play a pivotal role in mitigating this burden. Crops that have been genetically modified are often more resistant to disease and require a reduced level of insect control. Therefore, fewer acres need to be planted to reap the same amount of food.

One of the key players in the effort to increase the production of rice in Asia has been the Rockefeller Foundation. The foundation believes that GMOs will enable Asian farmers to obtain more rice from the same amount of land. "By the year 2005, the Foundation's hope is to increase rice production in Asia by 20 percent through the use of biotechnology without degrading the environment or reducing farm incomes." The Rockefeller Foundation has also been conducting similar work in Africa. Former President Jimmy Carter, who has become well known for his humanitarian works, has observed that "responsible biotechnology is not the enemy; starvation is. Without adequate food supplies at affordable prices, we cannot expect world health or peace".

Earth Save International disagrees with the notion that genetic engineering may play an important role in feeding the world's growing population. "World hunger is not a problem of technology or insufficient production, but primarily one of unequal distribution and economic inequality. As farmers lose their land and move to the cities, they also lose their food independence and begin to rely on money, often in dramatically short supply for many in the third world, in order to buy food that they formerly grew themselves".

In The Ecological Risks of Engineered Crops, Jane Rissler and Margaret Mellon agree. "Even if research is done on the right crops and result in increased agricultural productivity, increased production is only one factor in the complicated equation of world hunger. Poverty, trade policies, subsidies, soil erosion, and water shortages are also important causes of hunger. Increases in productivity obtained through genetic engineering will have mixed effects if not developed with due regard to the other important aspects of the hunger problem". In The Future of Food, Brian J. Ford, an English scientist, writer, and television host, says that a number of everyday farm animals and crops-Such as pigs, cattle, wheat, oats, barley, and rye-are the result of human intervention.

"Traditional farmers have been producing new animals and plants by cross-breeding for ten thousand years. " While genetic engineering "offers a more radical way of manipulating characteristics," it is fast becoming a part of daily life. "Genetic modification is inevitable. Like electric power, road transport, or computers, it is a facet of the future and the public will gain little by campaigning to ban this potentially rewarding technology. Properly applied, it could offer us so much". All the same, Ford does not believe that the public should simply accept whatever the bioengineering companies do. Rather, he calls for a careful monitoring of the industry and the introduction of safeguards. "This is a huge new industry, and it will have pronounced effects on us all.... We will need to control it". Ford suggests the following controls:

- Approval for new experiments should always be sought from a regulatory authority well versed in the subject.
- None of the members of such a committee should be in a position to benefit commercially from approval.
- No genes conferring problematical properties (e.g., antibiotic resistance) shall be used, whether as markers or otherwise, outside enclosed laboratories.
- Agents capable of transfer to wild plants, like pollen-carrying genes conferring resistance, shall not be liberated into the environment.
- The public shall be consulted about the siting and the benefits of experiments.

- All GMO products should be properly labeled and the source declared.
- Records of possible unwanted side effects should be meticulously maintained. All such events should be investigated by an organization unconnected with the source of the product.

ARGUMENTS AGAINST GMOS

The controversy shows no sign of diminishing. GMO opponents present some compelling arguments. The Organic Consumers Association contends that genetic engineering deconstructs basic, fundamental genetic barriers between species. "By combining the genes of dissimilar and unrelated species, permanently altering their genetic codes, novel organisms are created that will pass the genetic changes onto their offspring through heredity. Animal genes and even human genes are being inserted into plants or animals creating unimagined transgenic life". Completely new organisms will be created. The association believes that these have the potential to compromise human health and wellbeing, animal welfare, and the environment. The following are a few examples of ways in which this might occur:

- Genetically engineered organisms can reproduce, migrate, and mutate. If they escape or are released from the laboratory, they could wreak environmental havoc.
- Gene splicing can have unpredictable and dangerous results.
- Genetic engineering of crops and food-producing animals can produce toxic and allergic reactions in humans.
- Safety testing and regulation of genetically engineered organisms are inadequate or lacking.
- Patenting genes they discover and living organisms they create will enable a small corporate elite to own and control the genetic heritage of the planet.

The poll found that only about one-third of the Americans surveyed view GMOs as safe to eat. A little more than half found them unsafe for human consumption. Women, who tend to do the family shopping, were even more against GMOs than men. While 40% of men think that GMOs are unsafe, 62% of women feel that way. A striking percentage of those surveyed 93%-want the federal government to require food labels to indicate the presence of GMOs. Fifty-seven percent said that they were less likely to purchase food that contains GMOs. There are even age differences. "People under 45 are about 10 points more likely than their elders to think genetically modified foods are safe to eat. But a majority of young adults still calls genetically modified foods unsafe".

In Brave New Seeds, Robert Ali Brae de la Perriere, an international consultant, and Franck Seuret, a journalist, contend that throughout the world, GMOs will weaken the role of small-scale farmers who lack the wherewithal to pay higher prices for genetically modified seeds and lack the expertise that more technically oriented farming demands. Increasingly, large farms will absorb the smaller ones. Further, farmers are losing control over their products. "With transgenic plants, the freedom of agricultural practice and liberty of choice is reduced as the farmer works with a patented product which is subject to very specific conditions of use. For example, contracts for the use of a transgenic soya variety that is resistant to the herbicide RoundUp forbids, amongst other things, the cultivation of other varieties, the use of herbicides other than those allowed by Monsanto and the exchange of seeds with neighbours".

Crop diversification reduces the amount of herbicide and synthetic nitrogen fertilizer that may be needed, improves the quality of soil and water, controls insect-pest and pathogen populations, increases crop yield, and reduces yield variance. Insect-resistant crops may have unintended victims and disrupt other forms of pest control. Also, toxins from these products may remain in the soil for extended periods of time and reduce the level of soil fertility.

As if this were not sufficiently problematic, there is the issue of "terminator" seeds, genetically altered plants that are programmed to kill their own seeds. Unlike most plants, these do not regenerate. "After the seeds are planted and the crop matures, the [seed-killing] toxin is produced, killing the new seeds the plants carry". In the previously noted Genetically Engineered Food, Martin Teitel and Kimberly A. Wilson say that terminator seeds end the long-held practice of saving seeds for future plantings. "Farmers who use terminator seeds will be forced to buy new seeds for every planting, ending an age-old tradition of seed saving and creating a perpetual cycle of dependence on big seed companies".

In Brave New Seeds, Brae de la Perriere and Seuret observe that terminator seeds "achieved their supreme goal enslaving their clients". There is an even more alarming possibility. "Some doomsday scenarios suggest pollen from Terminator plants could drift with the wind like a toxic cloud, cross with ordinary crops or wild plants, and spread from species to species until flora all around the world had been suddenly and irreversibly sterilized". That is probably a huge exaggeration. Gene drift does not occur that easily. A large number of crops pollinate themselves. As for those that do not, border fields may be established to contain genes.

In "Genetically Modified Foods: Are They a Risk to Human/Animal Health?," Generally, when testing products, the industry has compared a GM food to a non-GM food. "When they are not significantly different the two are regarded as 'substantially equivalent, and therefore the GM food crop is regarded as safe as its conventional counterpart." Pusztai said that "substantial equivalence is an unscientific concept that has never been properly defined and there are no legally binding rules on how to establish it." Additionally, there are currently no reliable methods to test new genetically engineered crops for food allergies. "It is at present impossible to definitely establish whether a new GM crop is allergenic or not before its release into the human/animal food/feed chain".

CHAPTER 8

Food Irradiation

Every year millions of people become ill from the food they eat. One way to improve the safety of food is to treat it with radiation. Many are opposed because they fear health risks. In 1993, two sisters, ages four and two, were eager to visit the newly built fast-food restaurant that had opened in their Seattle neighborhood. Early in January, their father treated them to cheeseburgers, fries, and milk. Six days later, the older child, who was suffering from cramps and bloody diarrhea, was rushed to the hospital. Soon her sister was also hospitalized. Both girls were horribly ill, and during the course of her sickness, the older child often teetered close to death.

Their illness was caused by food poisoning. The hamburger meat in the cheeseburgers was contaminated with Escherichia coli 0157:H7, a deadly bacterium. Fortunately, both sisters survived their ordeal, but others did not. During that West Coast outbreak, several children died.

Food is supposed to nourish and sustain us. On occasion, food is improperly handled. When that occurs, deadly bacteria are provided an ideal medium in which to flourish. Statistics from the Centers for Disease Control and Prevention (CDC) have estimated that every year between 6 and 33 million people in the United States experience a foodborne disease. Of these, about 9,000 die. That means that on the average, every week slightly fewer than 200 people in the United States die from something they have eaten. While anyone is vulnerable, the young and the old or those with impaired immune systems are most at risk for serious morbidity or death.

Salmonella and Campylobacter are two of the most frequent pathogens that cause foodborne illness. "The Centers for Disease Control and Prevention estimates that Salmonella-commonly found in poultry, eggs, meat and milk-sickens as many as four million and kills 1,000 per year nationwide. Campylobacter, found mostly in poultry, is responsible for six million illnesses and 75 deaths per year in the United States," and "nationally, E. coli 0157:H7 causes about 20,000 illnesses and 500 deaths a year".

The financial toll is also high. "In 1996, the estimated cost of direct medical care and lost productivity resulting from all intestinal and foodborne infectious disease was about $28 billion". Companies lose money because they have to buy back contaminated products and pay huge settlements to affected families.

The CDC contends that about 97% of all food-related illnesses are a direct result of improper handling. Of these, 79% occur in a commercial setting; 21% take place at home. Not everyone is aware of or understands the concept of improperly handled food. Food can be inappropriately handled in a number of ways. It may be inadequately cooked or may be allowed to remain too long at a temperature that fosters bacterial growth. Also, unless people preparing the food frequently wash their hands, most notably after using the restroom, they have the potential to spread germs, especially if they are ill.

Further, there is the problem of cross contamination. "Cross-contamination may occur when raw contaminated food comes in contact with other foods, especially cooked foods through direct contact or indirect contact on food preparation surfaces". Ground beef is particularly risky "because the grinding process spreads any pathogens that may be present on the surface of the meat throughout the ground beef, " and because "an individual hamburger patty may contain meat from many cattle", thereby increasing the number of sources of contaminants.

Clearly, these figures could be lowered somewhat if people were more informed. Some believe that a far broader effort must be made. They call for the commercial irradiation of many types of food, a process that exposes food to radiation and is known to minimize the number of pathogens as well as extend the useful life of perishable foods. Irradiation is already used to reduce the bacteria in hospital supplies, medical instruments, wine-bottle corks, and cosmetics. Since 1972, American astronauts have eaten foods that have been irradiated. "Some hospitals and nursing homes serve irradiated food to people with compromised immune systems, including burn victims and those undergoing chemotherapy".

Since radiation evokes images of atomic explosions and nuclear war, it is often thought of as being dangerous. As might be imagined, some people strongly support the irradiation of food, while others just as vehemently oppose it. The vast majority of people are poorly informed about irradiation. A Food Marketing Institute telepHone survey in 1996 found that 47% of the 1,007 respondents knew nothing about food irradiation. In The Food That Would Last Forever, Gary Gibbs wrote, "If you were to ask the average person on the street what he or she thinks of food irradiation, the most likely response would be a puzzled, 'What is that?' Few Americans know that the next time they bite into a slice of bread, they may be eating irradiated wheat, or that when they pour spaghetti sauce on their pasta, they may be eating irradiated herbs and spices".

HISTORY OF FOOD IRRADIATION

Irradiation has an interesting history. Shortly before the end of the nineteenth century, the French pHysicist Antoine-Henri Becquerel discovered radiation, or radiant energy. Soon thereafter, Samuel Prescott, a biology professor at the Massachusetts Institute of Technology in Cambridge, Massachusetts, demonstrated that bacteria in food could be killed with gamma rays from radium.

He suggested that radiation be used to help preserve a variety of foods, including meat, fruits, vegetables, and grains. "By 1930, patents had been granted for food irradiation in both the United States and France", but more extensive studies did not begin until the World War II era. In 1953, President Dwight D. Eisenhower's Atomic Energy Commission spearheaded the "Atoms for Peace" programme designed to find nonviolent means tp use atomic energy. Food irradiation, or the exposure of food to "very high-energy invisible lightwaves", was one such use of atoms. Other countries were also entering the field. "The Soviet Union led the way by permitting the irradiation of potatoes (to prevent sprouting) in 1958 and the irradiation of grain (for disinfestations) in 1959. West Germany, although the first to actually employ commercial food irradiation in 1957 for the sterilization of spices used in sausage, dropped back in the race in 1958, when it banned food irradiation for use on foods consumed within its own country".

The Food, Drug, and Cosmetic Act of 1958 defined food irradiation as an additive rather than a process, such as pasteurization. "In practical terms, this means that every single irradiated food, as well as most forms of packaging, must be approved by the FDA.

Soon the FDA was giving its approval to irradiation of various foods. In 1963, irradiation of wheat and flour was approved, followed by potatoes, dried spices and vegetable seasonings, pork, fruits and vegetables, and poultry during the next three decades.

FDA approved irradiation of red meat on December 2, 1997, after three years of study and "a series of high-profile food poisonings and meat recalls". Irradiated foods must be labeled either "treated with radiation" or "treated by irradiation" and must contain a picture of the radura, the international symbol of irradiation. These rules only affect food products sold in stores. "When used as ingredients in other foods... the label of the other food does not need to describe these ingredients as irradiated". Also, food need not be labeled if one of its minor ingredients, such as a spice, has been irradiated.

No one refutes the effectiveness of food irradiation, which is sometimes called cold pasteurization because it kills bacteria without heat. Radiation energy alters molecules and breaks chemical bonds. In so doing, it has the potential to retard spoilage, thus reducing the chance of human illness.

Not all foods are suitable candidates for irradiation. "Cucumbers, grapes and some tomatoes turn mushy when radiation breaks cell walls and release enzymes that digest the food and speed up rotting". Oysters and other raw shellfish may be irradiated, but the process damages or kills the live oyster, thus greatly diminishing its quality. Eggs are also not good candidates for irradiation. "Irradiation causes the egg whites to become milky and more liquid, which means it looks like an older egg, and may not serve as well in some recipes".

TYPES OF FOOD IRRADIATION

At present there are several main types of food irradiation. The most common variety uses cobalt 60 as its radiation source. A cobalt-based irradiator requires space for a deep pool of shielding water and is thus impractical for most food plants. "Processors who wish to irradiate their food with cobalt must ship it to the facility, which represents an ongoing expense and logistical problem".

The first such facility was built in the early 1990s in Mulberry, Florida. "The heart of the plant... is a shiny rack of 400 gamma-ray-emitting cobalt-60 'pencils,' each 18 inches long and the diameter of a fat crayon, housed in a chamber surrounded by a concrete wall six feet thick. When not in use, the rack is submerged in a 15-foot-deep pool of cooled water that absorbs and neutralizes the gamma rays. At the push of a button, hydraulic arms slide into the irradiation chamber on an overhead monorail. The boxes follow a zig-zag pattern around the radioactive rack so gamma rays can reach all sides. Treatment times vary-fresh strawberries pass through in five to eight minutes, frozen chicken takes as long as 20 minutes".

Cesium irradiators require much less space-only an 8- to 10-foot area-and use fuel from decommissioned nuclear weapons, "which makes for a cheaper supply than cobalt (although not as stable)". Because there are only a small number of nuclear-weapons-reprocessing facilities throughout the world, there are only a few cesium irradiators.

Electron-beam irradiators are safer for workers and the environment because they may be turned off. It is also easier to change their elements. Although they deliver faster and higher doses of radiation than most other sources, they lack penetrating power. "Radioactive source irradiators emit gamma rays, which consist of pHotons. Because pHotons are units of electromagnetic energy with no mass, they penetrate relatively well, which allows the irradiator to process entire pallets at a time. The electrons in electron beams, on the other hand, have mass, which greatly reduces their penetrating power. In practical terms, the average electron-beam irradiator can penetrate only two to four inches, meaning that only relatively flat products (like hamburger patties) and/or packages are suitable".

X-ray irradiators-more powerful versions of the types of machines used in hospitals and dental offices-are still in development. "A beam of electrons is directed at a thin plate of gold or other metal, producing a stream of X-rays coming out the other side. Like cobalt gamma rays, X-rays can pass through thick foods, and require heavy shielding for safety. However, like e-beams [electron beams], the machine can be switched on and off, and no radioactive substances are involved". In all forms of food irradiation, the food is irradiated after it has been packaged and sealed. Food that may have been tainted during any portion of the production process is cleaned.

Food irradiation is administered in one of three "doses." A dose is the amount of radiation absorbed by the food and is determined by the intensity of the radiation and the length of time it is applied. The following are examples of usages of treatment doses approved by the Food and Drug Administration (FDA): Low doses are used to control insects in grains, inhibit sprouting in white potatoes, control trichinae in pork, and inhibit decay and control insects in fruits and vegetables. Medium doses control salmonella, shigella, Campylobacter, yersinia, and other bacteria that can cause food poisoning in meat, poultry, arid fish, and delay mold growth on strawberries and other fruits. High doses kill micro-organisms and insects in spices and commercially sterilize food.

Doses of irradiation are measured in Gray (GY) units. "This is a measure of the amount of energy transferred to food, microbe or other substance being irradiated. A single chest X- ray has a dose of roughly a half of a milliGray. To kill Salmonella, fresh chicken can be irradiated at up to 4.5 kiloGrays, which is about seven million times more irradiation than a single chest X-ray. The killing effect of irradiation on microbes is measured in D-values. One D-value is the amount of irradiation needed to kill 90% of that organism.

THE CASE FOR IRRADIATION

Currently, almost 40 countries have approved food irradiation for more than 40 different foods- "from apples in China and frog legs in France to rice in Mexico, raw pork sausages in Thailand, and wheat in Canada". Throughout the world, there are about 160 irradiators.

"The plants, which must be approved by governmental authorities before construction, are subject to regular inspections, audits and other reviews to ensure that they are safely and properly operated". Further, food irradiation has been endorsed by a wide range of powerful and prestigious groups, including the World Health Organization, the United Nations, the International Atomic Energy Association, the U.S. Public Health Service, the Food and Drug Administration, the American Medical Association, the American Public Health Association, and the American Dietetic Association. Every year, "about a half million tons of food products and ingredients are irradiated worldwide".

The International Food Information Council, a Washington, DC-based nonprofit group that was formed to communicate information on food safety and nutrition, contends that years of research have proven that food irradiation is safe. "The first patents for food irradiation were granted in the United States and United Kingdom in 1905. Since then, hundreds of studies on the effects of food irradiation have been reviewed by the U.S. government and other governments worldwide. It is the most extensively studied food processing technology available".

The International Food Information Council says that irradiated food does not become radioactive. "The food irradiation process moves food through a radiant energy field, but the food never touches the energy source. Think of the irradiation process as the way light passes through a window".

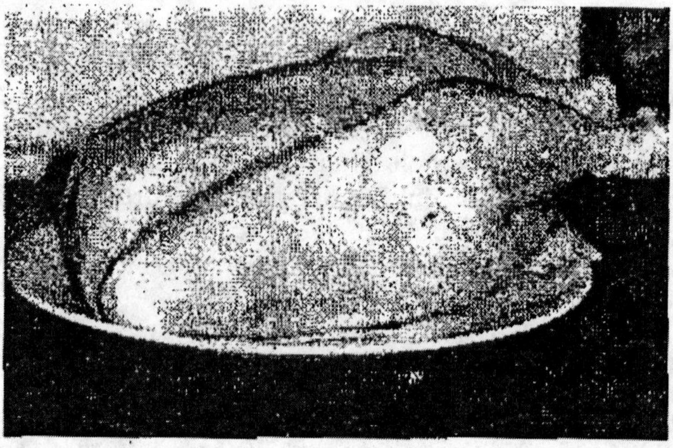

Fig. Irradiated Foods are Becoming more Commonplace.

It's similar to what happens when you get an x-ray. The rays pass through your body, but don't leave you radioactive". Also, the environment in which we live contains trace amounts of radioactivity. In the course of everyday life, people are exposed to small amounts of radiation from sources such as their television sets.

How are the nutrients and flavors of food altered by irradiation? To these questions the International Food Information Council responds, "Irradiation produces virtually no heat within food and does not 'cook' foods. Foods processed with irradiation are just as nutritious and flavorful as other food in the marketplace. Changes in nutrient content or flavor from cooking, canning, or freezing are the same or lower with processing by irradiation." The council maintains that irradiation greatly improves the safety of food, but cautions that the food is still subject to spoilage or to contamination during food handling and preparation. The process of irradiating food does affect its chemical composition. Could some of these changes be harmful? The National Food Safety Database, which is supported by the Department of Agriculture, comments that "in general, the irradiation process produces very little chemical change in food. None of the changes known to occur have been found to be harmful or dangerous".

How about free radicals, which are known to be released during food irradiation? "Free radicals-which in scientific terms are atoms or molecules with an unpaired electron-can be formed during the irradiation process, as well as by certain other food treatments (toasting bread, frying and freeze-drying) and during normal oxidation processes in food." There is absolutely no proof that free radicals alter the safety of irradiated foods. "Free radicals disappear by reacting with each other in the presence of liquids such as saliva in the mouth. Consequently, their ingestion does not create any toxicological or other harmful effects."

In a long-term laboratory study in which animals were fed dry milk powder irradiated by 45 kilogray, more than four times the maximum dose approved for food irradiation, "no mutagenic effects were noted and no tumors were formed. No toxic effects were apparent in the animals over nine successive generations. Similarly, toasted bread (unirradiated), which actually contains more free radicals than very dry foods that have been irradiated, can be expected to be harmless". Can food irradiation trigger genetic abnormalities? In the 1970s, studies conducted in India reported higher rates of multiple sets of chromosomes in rats, mice, monkeys, and malnourished children who had eaten wheat immediately after irradiation. No such results were found in subjects who had eaten wheat 12 weeks after irradiation.

Since then, no other institution has replicated these results. Several years later, in the early 1980s, human volunteers in China were fed irradiated food, including irradiated wheat. In eight different studies, more than 400 people ate irradiated food for 7 to 15 weeks. "No significant difference between the number of chromosomal aberrations in the control and test groups could be discovered." Many subsequent studies have supported this finding. "Over the last 20 years millions of mice, rats and other laboratory animals have been bred and reared exclusively on an irradiated diet. No transmittable genetic defects have been observed which could be attributed to the consumption of irradiated diets". There are other safety issues. Accidents have occurred at food-irradiation facilities, and people who work there are at risk from radiation-related accidents, but the National Food Safety Database believes that such problems are rare: "Over the past 25 years. there have been a few major accidents at industrial irradiation facilities that caused injury or death to workers because of accidental exposure to a lethal dose of radiation. All of the accidents happened because safety systems had been deliberately bypassed and proper control procedures had not been followed. None of these accidents endangered public health and environmental safety."

Accidents can also occur during the transportation of radioactive materials. These concerns should be "weighed against the fact that in all the decades medical equipment and various foods have undergone irradiation, there has never been a nuclear accident attributable to irradiation materials". Moreover, "in the 50 or so years that medical supplies and various foods have undergone irradiation, there has not been a single nuclear accident in which a finger could be pointed at irradiation materials".

Some worry that food irradiation will dramatically raise the price of food. Building a food-irradiation plant costs between one and three million dollars. "This is within the range of plant costs for other food technologies. For example, a moderately-sized, ultra-high temperature plant for sterilizing milk, fruit juices and other liquids costs about US $2 million." There are also costs associated with running the plants and irradiating the food. "Broken down, irradiation costs range from US $10 to $15 per ton for a low-dose application (for example, to inhibit the growth of sprouts in potatoes and onions) to US $100 to $250 per ton for a high-does application (for example, to ensure hygienic quality of spices). These costs are competitive with alternative treatments. In some cases, irradiation can be considerably less expensive".

Some stores and restaurants have taken the plunge. One of the first was Carrot Top, a grocery market in the Chicago area that started selling irradiated products in 1992. Jim Corrigan, the store's owner, has found that the food, which sells well, remains fresher for a far longer period of time. Purchased 50,000 pounds of irradiated chicken in 1997. At the time, the executive chef of the complex noted that regular chicken had to be discarded after five days of refrigeration. The irradiated chicken remained fresh for three weeks.

Many consumers seem willing to try such foods. A few years ago, the International Food Information Council paid for a series of focus groups to be conducted in four U.S. cities. "Participants from each group represented a diverse mix of consumers by age, gender, ethnic group, occupation, and education." The following were some of the findings:

- Most consumers do not fully understand and have not thought much about food irradiation as a processing or food-safety technology.
- After education, consumers were willing to try irradiated foods for themselves and their families, including children; however, for irradiated foods, taste was more important to consumers than price.
- Consumers embraced the food-safety benefits of food irradiation, but not as a substitute for safe food production, processing, and handling.
- When asked, unaided by text, about their understanding of the radura symbol that is used for identifying products treated with irradiation, most consumers thought the symbol was "cute, " pleasant, or appealing.
- The FDA currently requires irradiated foods to include the radura on the label accompanied by the statement "treated with irradiation." Consumers wanted food products to identify the process, tell what the process does, and describe public health benefits.
- Consumers felt that terms such as "cold pasteurized (irradiated) to eliminate harmful bacteria" provided the most beneficial information.
- Consumers appreciated the safety benefit of irradiation "to eliminate harmful bacteria" more than extended shelf-life. Consumers did not want produce that would "last for several months."

The Centers for Disease Control and Prevention contends that if consumers were better informed about the health benefits of food irradiation, they would be supportive. "Consumers are interested in a process that eliminates harmful microbes from the food and reduces the risk of food borne disease. In test marketing of specific irradiated foods, consumers have shown that they are willing to buy them. Typically at least half will buy the irradiated food, if given a choice between irradiated product and the same product non-irradiated. If consumers are first educated about what irradiation is and why it is done, approximately 80% will buy the product in these marketing tests".

David A. Kessler, dean of the Yale University School of Medicine and a former FDA commissioner, has compared the past debate over the pasteurization of milk to the current dispute around food irradiation. Although many people became sick from drinking milk, they remained skeptical about pasteurization. In 1889, Dr. Henry Koplik of New York City began an effort to make pasteurization a universally accepted practice. "The true point, however, in the public awareness and acceptance of pasteurization was when Nathan Straus, head of R.H. Macy's in New York City, launched a campaign to change the public's idea that pasteurized milk was a medicated product. Although Mr. Straus commended Dr. Koplik's work, he wanted a wider appeal for pasteurization and opened his first milk depot in 1893".

Straus's effort kept growing. Soon he was writing the mayors of the country's largest cities, and milk depots were established in many cities, including Boston, Chicago, Cleveland, PhiladelpHia, and Pittsburgh. "In 1895, Nathan Straus predicted that 'the day is not far distant when it will be regarded as a piece of criminal neglect to feed young children with milk that has not been sterilized (pasteurized).'" Kessler concludes that "though the ideas of men would have to be altered and the industry would have to develop to handle large-scale commercial pasteurization before that propHecy could come true, it would come true. Today we recognize milk pasteurization as a necessary safety measure".

THE OPPOSITION TO FOOD IRRADIATION

Despite widespread support, food irradiation is opposed by many. One of the more vocal opponents is Public Citizen, a watchdog organization founded in 1971 by consumer advocate RalpH Nader. Public Citizen says that there is really little evidence that food irradiation is safe. "Irradiation changes the molecular structure of food. Irradiation creates new and potentially dangerous compounds, called radiolytic products, by smashing apart the chemical bonds in food and sending electrons flying." Public Citizen states that there have been no long-term studies proving that food irradiation is safe. "In fact, almost all research on toxicity was done before 1980 and did not use modern toxicological methods of research".

Public Citizen maintains that food radiation reduces the nutritional value of food by destroying vitamins, essential amino acids, and polyunsaturated fatty acids. The longer shelf life of irradiated food can lead to even more loss of nutrients.

Public Citizen says that the FDA has chosen to ignore studies indicating that food irradiation may be dangerous and that it has caused health problems in laboratory animals such as shorter life spans, low birth weight, kidney damage, immune and reproductive problems, chromosomal abnormalities, and tumors. Unfortunately, since the FDA claims that much of this research was poorly done, the results were ignored. "Without exception, FDA officials have chosen to ignore any evidence that suggests irradiation may be dangerous". To Public Citizen, there is little value in comparing the radiation from televisions and microwaves to the amount of radiation used on food. "It would take 1.4 billion TVs to generate the amount of radiation generated when meat is 'treated' by an e-beam accelerator. In the case of microwave ovens, the devices use non-ionizing radiation, meaning that the radiation is not powerful enough to obliterate the chemical bonds that hold atoms and molecules together (which is also the case with visible and ultraviolet light)."

Although food-irradiation advocates claim that the food never becomes radioactive, Public Citizen says that it may, at least for a brief period of time. Moreover, any comparison between food irradiation and pasteurization is absurd. "There is nothing proper about comparing a process by which milk is quickly heated and cooled to a process by which food is exposed to the equivalent of tens of millions of X-rays-more than enough to turn nutritious, good-tasting food into vitamin-depleted, chemically altered mush that can smell like a wet dog". The Organic Consumers Association has also taken a strong stand against food irradiation. The organization notes that there is no definitive long-term evidence that irradiated food is safe:

- The longest human feeding study was 15 weeks. No one knows the long-term effects of a lifelong diet that includes foods that are frequently irradiated, such as meat, chicken, vegetables, fruits, salads, sprouts, and juices.
- There are no studies on the effects of feeding babies or children diets containing irradiated foods, except a small study from India.
- Studies on animals fed irradiated foods have shown increased tumors, reproductive failures, and kidney damage. Some possible causes are irradiation- induced vitamin deficiencies, the inactivity of enzymes in the food, DNA damage, and toxic radiolytic products in the food.
- The FDA based its approval of irradiation on only 5 of 441 animal-feeding studies. Marcia van Gemert, the toxicologist who chaired the FDA committee that approved irradiation, later said, "These studies reviewed in the 1982 literature from the FDA were not adequate by 1982 standards, and are even less accurate by 1993 standards to evaluate the safety of any product, especially a food product such as irradiated food." The 5 studies were conducted using lower energies of irradiation than those approved for foods for human consumption.
- The science of today is totally inadequate to prove the long-term safety of food irradiation.

The Organic Consumers Association believes that food irradiation is an environmental hazard. If food irradiation grows in popularity, then the likelihood of accidents-whether in transport, operation, or disposal of nuclear wastes-will increase. There have already been problems. "In Georgia in 1988, radioactive water escaped from an irradiation facility. The taxpayers were stuck with $47 million in cleanup costs. Radioactivity was tracked into cars and homes. Radioactive water was flushed into the public sewer system. " "Numerous worker exposures have occurred in food irradiation facilities worldwide". Further, while food irradiation does reduce the amount of bacteria in food, the Organic Consumers Association empHasizes that it does not ensure clean food. Consider the following:

- Bacteria that survive irradiation are by definition radiation-resistant. These bacteria will multiply and eventually work their way back to the slaughterhouses, perhaps through a slaughterhouse worker who ate irradiated meat for dinner. The bacteria that contaminate the meat will no longer be killed by currently approved doses of radiation, and stronger bacteria will contaminate our food supply.
- People may become more careless about sanitation if irradiation is widely Used. In a few hours at room temperature, the bacteria remaining in meat or poultry after irradiation can multiply to the level existing before irradiation. Some bacteria (like the one that causes botulism), viruses, and prions (which are believed to cause mad cow disease) are not killed by current doses of irradiation.
- Irradiation encourages food producers to cut corners on sanitation, because they can irradiate the food just before it is shipped.

A third leader in the anti-food-irradiation movement is Food and Water, a national nonprofit food-safety and environmental organization based in Walden, Vermont. Food and Water states that the government is approaching the problem of foodborne illness from the wrong direction and is failing to address the primary reasons for the U.S. food-safety crisis. Instead of advocating the irradiation of meat, the government should implement immediate steps to clean up the current system of meat production. Food and Water identifies the following causes of dirty meat:

- *Factory farms:* Factory farms and feed lots are the initial breeding ground of pathogens that contaminate the food supply. Cattle are frequently confined to inhumane pens, open and unprotected from weather, often in groups of 500 or more. Before even entering the slaughterhouse, cattle can be contaminated by feed, water, feces, and other animals.
- *Slaughtering facilities:* Industrial slaughtering facilities are fast, filthy, and dangerous. Many large slaughtering facilities operate at rates that exceed 300 cows per hour. These breakneck speeds are the reason that jobs in these facilities are now among the most dangerous in the country. At these high rates, filth and the spread of pathogen-containing fecal matter to consumable meat become commonplace.
- *Industrial food:* The nation's love affair with cheap, industrial, "fast, " and processed food creates the demand that eventually leads to factory farming, an empHasis on consumption and devaluation of production, and a disconnection between citizens and their food sources. Until we

come to grips with the fact that cheap industrial food is inherently unsafe (irradiated or not), we can only expect things to get worse.

- *Meat monopolies:* A mere three corporations control 77% of the meat market, giving them a monopoly-like control over beef production, slaughtering, and distribution. This kind of concentration translates to contamination as long as speed and profit, rather than health and public safety, remain the priorities of the meat industry.

In Food Irradiation: An Activist Primer, Food and Water includes quotes opposing food irradiation from such scientists as George L. Tritsch of the Roswell Park Cancer Institute in Buffalo, New York, John W. Gofman of the University of California at Berkeley, and Geraldine Dettman, radiation safety officer and biosafety officer at Brown University. These scientists raise concerns about carcinogens, chemicals, and mutated organisms produced by irradiation.

CHAPTER 9

Fermentation

THE BENEFITS OF FERMENTING FRUITS AND VEGETABLES

Fermenting fruits and vegetables can bring many benefits to people in developing countries. Fermented foods play an important role in providing food security, enhancing livelihoods and improving the nutrition and social well being of millions of people around the world, particularly the marginalised and vulnerable.

IMPROVING FOOD SECURITY

Eight hundred million people do not have enough food to eat. If we include those not free from hunger the figure rises to 1. 2 billion people. This is one fifth of the World's population. A further two billion people are deficient in one or more micronutrients. In the seventies, food security was viewed mainly in terms of food supply at the global and national levels. Since then there has been a major shift in understanding of food security with more empHasis on access to food rather than purely on production.

The Food and Agriculture Organisation of the United Nations (FAO), amongst other influential organisations, has recognised that the problem of food security cannot be tackled in isolation. Moreover that it is an integral component of other development issues. FAO highlights the fact that the world food insecurity problem is a result of undemocratic and inequitable distribution of and access to resources rather than a problem of global food production.

Fermentation technologies play an important role in ensuring the food security of millions of people around the world, particularly marginalised and vulnerable groups. This is achieved through improved food preservation, increasing the range of raw materials that can be used to produce edible food products and removing anti- nutritional factors to make food safe to eat.

FOOD PRESERVATION

Fermentation is a cheap and energy efficient means of preserving perishable raw materials. When harvested, fruit and vegetables, undergo rapid deterioration, especially in the humid tropics where the prevailing environmental conditions accelerate the process of decomposition. There are several options for preserving fresh fruit and vegetables including drying, freezing, canning and pickling. However many of these are inappropriate for use on the small-scale in developing countries. For instance the canning of vegetables at the small- scale has serious food safety implications and contamination with botulism is a possibility.

Freezing of fruits and vegetables is not economically viable at the small-scale. Fermentation requires very little sopHisticated equipment, either to carry out the fermentation or for subsequent storage of the fermented product. It is a technique that has been employed for generations to preserve food for consumption at a later date and to improve food security. There are examples from around the world of the role fermented foods have played in preserving food to enhance food security.

Fermented Foods for Survival in Sudan

About 60% of the fermented foods of Sudan are famine or survival foods. Many of the fermented foods have been developed in Western Sudan in the Kordofan and Darfur regions, which are traditional famine areas. The strong link between fermented foods and food shortages is revealed by the fact that when a family becomes rich a number of fermented foods are no longer prepared. The techniques used are very effective methods of food preservation. The products can be preserved for years through the double action of fermentation itself (which produces anti-microbial acids) and sun-drying. Sudan is probably the hottest and driest country in Africa.

Through the years women have made full use of this free solar energy. Shade temperatures in the summer reach 45-50°C and the hot sands outside the shade reach more than 70°C. Dried and fermented foods together with the seeds and fruits that can be gathered from the wild have saved lives especially those of children in the past and in the present in times of shortage. During the 1983-85 famine, relief workers found that people had survived by producing specific traditional fermented food products, especially Kawal.

Gundruk: an Important Fermented Product in Nepal

Gundruk is a fermented and dried vegetable product. It is produced by shredding the leaves of mustard, radish and cauliflower leaves and placing them in an earthenware pot to ferment. After five to seven days the leaves are removed and dried in the sun. Gundruk is a very important food product in Nepal ensuring food security for many Nepali communities especially in remote areas.

It is served as a side dish with the main meal and is also used as an appetiser in the bland, starchy diet. The annual production of gundruk in Nepal is estimated at 2,000 tons and most of the production is carried out at the household level.

Gundruk is also an important source of minerals particularly during the off-season when the diet consists of mostly starchy tubers and maize which tend to be low in minerals.

SALVAGING WASTE FOODS

Fermentation can salvage waste food which otherwise would not be usable as food by changing the consistency of the product and making it digestible. This increases the range of raw materials available as food.

Bones and Hides

A wide range of "waste" products are fermented to produce edible food products in Sudan. These includes bones, hides and locusts. Fresh bones are fermented into a variety of products. "Dodery" is produced by chopping bones into small pieces and placing them into fermenting vats. They are subsequently covered in water, left for three days, removed, crushed into a paste and mixed with the ash from burnt sorghum stalks.

The mixture is returned to the fermenting vat for a further two to five days. The final product is rolled into balls and has a shelf like of up to two months. Another product "Kaidu digla" is made from the vertebrae of the backbone. These are chopped into smaller pieces and then sun-dried. After drying they are pounded with stones; mixed with water and salt; moulded into balls and allowed to ferment.

Use of Waste Products in Indonesia

In Indonesia a variety of waste products are fermented to produce nutritious food products. Tempe-bongrek is a protein rich food made in Indonesia by fermenting peanut and coconut press-cake, remaining after oil extraction. The product is similar to traditional tempeh produced from the fermentation of soya beans.

The production of tempeh- bongrek is a mould fermentation, initiated by inoculation of the soaked, acidified press-cake with Rhizopus species. The inoculated cakes are placed on banana leaves or plastic sheets in a dark room for about 2 days.

An incubation temperature of 37°C is optimal for the mould and prevents the growth of P. cocovenenans, which produces bongkrek toxin. A pH of less than 6.0 also prevents the development of bongkrek toxin. Ontjom is produced from waste groundnut press cake, tapioca waste and the solid waste of tahu.

Ontjom is prepared using a mixed culture of microorganisms with Rhizopus or Neurospora species predominating. Ontjom is mainly produced in west Java where it is consumed as a side dish in the form of deep fried slices. It forms an important daily food item for the west Javanese, particularly those from the lower income groups. Fresh coconut residue, left over from the production of coconut cream or milk, can be fermented by Bacillus subtilis, in an alkaline fermentation, to produce semayi, which is widely consumed in Indonesia.

Pineapple Peel Vinegar

A considerable amount of the fruit can be wasted in the peeling and preparation of pineapples in Latin America. Through fermentation, a product can be produced from the peelings, that would otherwise have been discarded. The peelings are placed in containers of water and sugar and yeast are added. They are allowed to sit for about 8 days, after which a pineapple peel vinegar is produced. The product is of a distinct, light pineapple flavour and can be used in the same way as other vinegars.

REMOVAL OF ANTI-NUTRITIONAL FACTORS

Many fruits and vegetables contain naturally occurring toxins and anti-nutritional compounds. These can be removed or detoxified by the action of micro-organisms during fermentation. For instance the fermentation process that produces the Sudanese product Kawal removes the toxins from the leaves of Cassia obtusifolia and fermentation is an important step in ensuring that cassava is safe to eat.

Removing Cyanide by Fermentation

Cassava contains a naturally occurring chemical: cyanogenic glucoside. When eaten raw or improperly processed, this substance releases cyanide into the body, which can be fatal. Correct processing removes this chemical. The cassava is first peeled (as about 60-70% of the poison is in the peel) and then soaked in stagnant water or fermented in sacks for about three days. It is sometimes grated or rasped as this helps to speed up the fermentation process. At the beginning of the fermentation, Geotricum Candida acts on the cassava. This tends to make the product acidic, which finally kills off the micro-organisms as they cannot exist in such a medium.

A second strain of micro-organisms (Cornibacterium lactii), which can tolerate the acidic environment then take over and by the third day 90-95% of the dangerous chemical will have been hydrolysed. The cassava also develops its characteristic flavour. The product is then sieved and the fine starch particles are fried in an iron pan alone over a flame or with some palm oil. During this process most, if not all the remaining toxins are given off. The liquor from a previous fermentation is used as a starter, thereby reducing the period of fermentation to about 6-8 hours.

INCREASING INCOME AND EMPLOYMENT

The production of fermented fruit and vegetable products provides income and employment to millions of people around the world. Food processing is probably the most important source of income and employment in Africa, Asia and Latin America. The Food and Agriculture Organisation of the United Nations has stated that value added through marketing and processing raw products can be much greater than the value of primary production. For instance in sub-Saharan

Africa more than 60% of the workforce is employed in the small scale food processing sector, and between one third and two thirds of value added manufacturing is based on agricultural raw materials. This is particularly important as agriculture and the formal sector are unable to absorb the growing labour force in many countries. Fermented foods are popular throughout the world and the production of fermented food products is important in many countries in providing income and employment.

In Africa, fermented cassava products (like Gari and Fufu) are a major component of the diet of more than 800 million people and in some parts of Africa it constitutes over 50% of the diet. In Asia the preparation of fermented foods is a widespread tradition. Kimchi (a fermented cabbage product) is the major food product of Korea. Soy sauce (a fermented legume product) is economically important from Indonesia to Japan. Over a billion litres are produced each year in Japan alone. Over 2000 million litres are produced each year in Korea and over 150 million litres in Taiwan. Miso (a fermented legume product) is also very important in Asia with over 560, 000 tons produced *a* year in Japan alone. In Latin America, fermented cereal products, alcoholic drinks and fermented milk products are three of the most important sectors of the economy.

IMPROVING NUTRITION

The optimum health and nutrition of individuals is dependent upon a regular supply of food and a balanced diet When diets are sub-optimal, the individual's capacity for work and achievements are greatly reduced. The most vulnerable groups are women, children and weaning infants. Availability of food, dietary restrictions and taboos, misconceptions, limited time available for feeding or eating compound to create a group of individuals who are nutritionally disadvantaged. Approximately 30% of women consume less than their daily requirements of energy and at least 40% of women world-wide suffer from iron-deficiency anaemia. Fermentation can enhance the nutritional value of a food product though increased vitamin levels and improved digestibility.

VITAMINS

Fermentation processes can result in increased levels of vitamins in the final product. Saccharomyces cerevisiae is able to concentrate large quantities of thiamin, nicotinic acid and biotin and thus form enriched products.

- Sorghum beer in Southern Africa contains relatively high levels of riboflavin and nicotinic acid, which are important for people consuming a high maize diet. Pellagra (a vitamin deficiency disease associated with high maize diets) is unusual in communities in which sorghum beer is consumed. Even children benefit from consuming the dregs which contain relatively little alcohol but are rich in vitamins.

- Palm wine in West Africa is high in vitamin B12, which is very important for people with low meat intake, and who subsist primarily on a vegetarian diet.
- Pulque (a fermented plant sap) is an important source of vitamins for the economically deprived in Mexico. The fermentation process involved in Pulque production increases its vitamin content. For instance the vitamin content (milligrams of vitamins per 100g of product) of pulque increases from 5 to 29 for thiamine, 54 to 515 for niacin and 18 to 33 for riboflavin during fermentation.
- Idli (a lactic acid bacteria fermented product consumed in India) is high in thiamine and riboflavin.

DIGESTIBILITY

Micro-organisms contain certain enzymes, such as cellulases, which are incapable of being synthesised by humans. Microbial cellulases hydrolyse cellulose into sugars which are then readily digestible by humans. Similarly pectinases soften the texture of foods and liberates sugars for digestion. Fermented foods are often more easily digestible than unfermented foods.

Lactic acid fermented weaning foods are traditionally produced in developing countries, both to improve the safety of the food and to improve its digestibility. Starchy porridges are commonly fed to weaning infants in developing countries. The consistency of these gruels, combined with the small capacity of the infants stomach, means that it is physically impossible for the child to consume adequate energy to meet its high demands. By acidifying the porridge through lactic acid fermentation, starch is hydrolysed into shorter chains of glucose and dextrose, which reduce the viscosity of the porridge and increase its energy density. Thus the child is more able to meet its energy requirements.

MEDICINAL BENEFITS

There are many traditional beliefs about the medicinal properties of fermented food products. The Fur ethnic group in Sudan strongly believe that the consumption of fermented foods protects them from disease. Koumiss (a fermented milk product in Russia) has been used to treat tuberculosis. Pulque (a fermented fruit sap) is felt to have medicinal properties in Mexico. There is a sound scientific basis to these assertions:

- The lowering of the pH inhibits the growth of food spoiling or poisoning bacteria and destroys certain pathogens.
- Certain lactic acid bacteria (e.g. Lactobacillus acidopHilus) and moulds have been found to produce antibiotics and bacteriocins.
- The beneficial health effects of lactic acid bacteria on the intestinal flora are well documented.

- Substances in fermented foods have been found to have a protective effect against the development of cancer.

Fermentation is a traditional method of reducing the microbial contamination of porridges in Kenya. A study in Tanzania has shown that children fed with fermented gruels had a 33% lower incidence of diarrhoea than those fed unfermented gruels, owing to the inhibition of pathogenic bacteria by lactic acid forming bacteria.

IMPROVING CULTURAL AND SOCIAL WELL BEING

Fermentation can improve the flavour and appearance of food. One important area is the creation of meat-like flavour. Over the years, Sudanese women have developed products to replace meat in their diets. These include "kawal", fermented wild legume leaves, "sigda" (fermented sesame press-cake) and "furundu" (fermented red sorrel seeds). The strong flavours of fermented food products can enhance a dull diet. Fermented vegetables such as pickles, gundruk and sauerkraut are used as condiments to enhance the overall flavour of the meal. A small amount of pickle can make a bland starchy diet (like dahl and rice in Asia) much more appealing.

BASIC PRINCIPLES OF FERMENTATION

THE DIVERSITY OF FERMENTED FOODS

Numerous fermented foods are consumed around the world. Each nation has its own types of fermented food, representing the staple diet and the raw ingredients available in that particular place. Although the products are well know to the individual, they may not be associated with fermentation. Indeed, it is likely that the methods of producing many of the worlds fermented foods are unknown and came about by chance. Some of the more obvious fermented fruit and vegetable products are the alcoholic beverages - beers and wines. However, several more fermented fruit and vegetable products arise from lactic acid fermentation and are extremely important in meeting the nutritional requirements of a large proportion of the worlds population.

ORGANISMS RESPONSIBLE FOR FOOD FERMENTATIONS

The most common groups of micro-organisms involved in food fermentations are:

- Bacteria
- Yeasts
- Moulds

Bacteria

Several bacterial families are present in foods, the majority of which are concerned with food spoilage. As a result, the important role of bacteria in the fermentation of foods is often overlooked. The most important bacteria in desirable food fermentations are the lactobacillaceae which have the ability to produce lactic acid from carbohydrates. Other important bacteria, especially in the fermentation of fruits and vegetables, are the acetic acid producing acetobacter species.

Yeasts

Yeasts and yeast-like fungi are widely distributed in nature. They are present in orchards and vineyards, in the air, the soil and in the intestinal tract of animals. Like bacteria and moulds, yeasts can have beneficial and non-beneficial effects in foods. The most beneficial yeasts in terms of desirable food fermentation are from the Saccharomyces family, especially S. cerevisiae.

Fermented Foods from Around the World.

Name and region	Type of product
Indian sub-continent	
Acar, Achar, Tandal achar, Garam nimboo achar	Pickled fruit and vegetables
Gundruk	Fermented dried vegetable
Lemon pickle, Lime pickle, Mango pickle	
South East Asia	
Asinan, Burong mangga, Dalok, Jeruk, Kiam-chai, Kiam-cheyi, Kong-chai, Naw-mai-dong, Pak- siam-dong, Paw-tsay, Phak-dong, Phonlami-dong, Sajur asin, Sambal tempo-jak, Santol, Si-sek-chai, Sunki, Tang-chai, Tempoyak, Vanilla,	Pickled fruit and vegetables
Bai-ming, Leppet-so, Miang	Fermented tea leaves
Nata de coco, Nata de pina	Fermented fruit juice
East Asia	
Bossam-kimchi, Chonggak-kimchi, Dan moogi, Dongchimi, Kachdoo kigactuki, Kakduggi, Kimchi, Mootsanji, Muchung-kimchi, Oigee, Oiji, Oiso baegi, Tongbaechu- kimchi, Tongkimchi, Totkal kimchi,	Fermented in brine
Cha-ts'ai, Hiroshimana, Jangagee, Nara senkei, Narazuke, Nozawana, Nukamiso-zuke, Omizuke, Pow tsai, Red in snow, Seokbakji, Shiozuke, Szechwan cabbage, Taitan tsoi, Takana, Takuan, Tsa Tzai, Tsu, Umeboshi, Wasabi-zuke, Yen tsai	Pickled fruit and vegetables
Hot pepper sauce	
Africa	
Fruit vinegar	Vinegar
Hot pepper sauce	
Lamoun makbouss, Mauoloh, Msir, Mslalla, Olive	Pickled fruit and vegetables

Name and region	Type of product
Oilseeds, Ogili, Ogiri, Hibiscus seed	Fermented fruit and vegetable seeds
Wines	Fermented fruits
Americas	
Cucumber pickles, Dill pickles, Olives, Sauerkraut,	Pickled fruit and vegetables
Lupin seed, Oilseeds,	Pickled oilseed
Vanilla, Wines	Fermented fruit and vegetable
Middle East	
Kushuk	Fermented fruit and vegetables
Lamoun makbouss, Mekhalel, Olives, Torshi, Tursu	Pickled fruit and vegetables
Wines	Fermented fruits
Europe and World	
Mushrooms, Yeast	Moulds
Olives, Sauerkohl, Sauerruben	Pickled fruit and vegetables
Grape vinegar, Wine vinegar	Vinegar
Wines, Citron	Fermented fruits

Yeasts are unicellular organisms that reproduce asexually by budding. In general, yeasts are larger than most bacteria. Yeasts play an important role in the food industry as they produce enzymes that favour desirable chemical reactions such as the leavening of bread and the production of alcohol and invert sugar. Moulds are also important organisms in the food industry, both as spoilers and preservers of foods. Certain moulds produce undesirable toxins and contribute to the spoilage of foods.

Moulds

The Aspergillus species are often responsible for undesirable changes in foods. These moulds are frequently found in foods and can tolerate high concentrations of salt and sugar. However, others impart characteristic flavours to foods and others produce enzymes, such as amylase for bread making. Moulds from the genus Penicillium are associated with the ripening and flavour of cheeses.

Moulds are aerobic and therefore require oxygen for growth. They also have the greatest array of enzymes, and can colonise and grow on most types of food. Moulds do not play a significant role in the desirable fermentation of fruit and vegetable products. When micro-organisms metabolise and grow they release by-products. In food fermentations the byproducts play a beneficial role in preserving and changing the texture and flavour of the food substrate.

For example, acetic acid is the by-product of the fermentations of some fruits. This acid not only affects the flavour of the final product, but more importantly has a preservative effect on the food.

For food fermentations, micro-organisms are often classified according to these by-products. The fermentation of milk to yoghurt involves a specific group of bacteria called the lactic acid bacteria (Lactobacillus species). This is a general name attributed to those bacteria which produce lactic acid as they grow. Acidic foods are less susceptible to spoilage than neutral or alkaline foods and hence the acid helps to preserve the product. Fermentations also result in a change in texture. In the case of milk, the acid causes the precipitation of milk protein to a solid curd.

Enzymes

The changes that occur during fermentation of foods are the result of enzymic activity. Enzymes are complex proteins produced by living cells to carry out specific biochemical reactions.

They are known as catalysts since their role is to initiate and control reactions, rather than being used in a reaction. Because they are proteinaceous in nature, they are sensitive to fluctuations in temperature, pH, moisture content, ionic strength and concentrations of substrate and inhibitors. Each enzyme has requirements at which it will operate most efficiently. Extremes of temperature and pH will denature the protein and destroy enzyme activity. Because they are so sensitive, enzymic reactions can easily be controlled by slight adjustments to temperature, pH or other reaction conditions.

In the food industry, enzymes have several roles - the liquefaction and saccharification of starch, the conversion of sugars and the modification of proteins. Microbial enzymes play a role in the fermentation of fruits and vegetables.

Nearly all food fermentations are the result of more than one micro-organism, either working together or in a sequence. For example, vinegar production is a joint effort between yeast and acetic acid forming bacteria. The yeast convert sugars to alcohol, which is the substrate required by the acetobacter to produce acetic acid.

Bacteria from different species and the various microorganisms - yeast and moulds -all have their own preferences for growing conditions, which are set within narrow limits. There are very few pure culture fermentations. An organism that initiates fermentation will grow there until it's by-products inhibit further growth and activity.

During this initial growth period, other organisms develop which are ready to take over when the conditions become intolerable for the former ones.

In general, growth will be initiated by bacteria, followed by yeasts and then moulds. There are definite reasons for this type of sequence.

The smaller micro-organisms are the ones that multiply and take up nutrients from the surrounding area most rapidly. Bacteria are the smallest of micro-organisms, followed by yeasts and moulds.

The smaller bacteria, such as Leuconostoc and Streptococcus grow and ferment more rapidly than their close relations and are therefore often the first species to colonise a substrate.

Micro-Organisms Commonly Found in Fermenting Fruit and Vegetables

Organism	Type	Optimum conditions	Reactions
Acetobacter genus A. aceti A. pasteurianus A. peroxydans	Aerobic rods	aw >=0. 9	Oxidise organic compounds (alcohol) to organic acids (acetic acid). Important in vinegar production.
Streptococcaceae Family	Gram positive cocci	Acid tolerant aw >=0. 9	
Streptococcus genus S. faecalis S. bovis S. thermopHilus			Homofermentativ e. Most common in dairy fermentations, but S. Faecalis is common in vegetable products. Tolerate salt and can grow in high pH media.
Leuconostoc genus L. mesenteroides L. dextranicum L, paramesenteroides L. oenos	Gram positive cocci		Heterofermentati ve. Produce lactic acid, plus acetic acid, ethanol and carbon dioxide from glucose. Small bacteria, therefore have an important role in initiating fermentations. L. oenos is often present in wine. It can utilise malic acid and other organic acids.
Pediococcus genus P. cerevisiae P. acidilactici P. pentosaceus			SapropHytic organisms found in fermenting vegetables, mashes, beer and wort Produce inactive lactic acid.

Organism	Type	Optimum conditions	Reactions
Lactobacillaceae Family	Gram positive rods. Non-motile	Acid tolerant aw > =0. 9	Metabolise sugars to lactic acid, acetic acid, ethyl alcohol and carbon dioxide.
Lactobacillus genus			The genus is split into two types - homo- and heterofermenters. SapropHytic organisms. Produce greater amounts of acid than the cocci
Homofermentative Lactobacillus spp. L. delbrueckii L. leichmannii L. plantarum L. lactis L. acidopHilus			Produce only lactic acid. L. plantarum important in fruit and vegetable fermentation. Tolerates high salt concentration.
Heterofermentative Spp. L. brevis L. fermentum L. buchneri			Produce lactic acid (50%) plus acetic acid (25%), ethyl alcohol and carbon dioxide. (25%). L. brevis is the most common. Widely distributed in plants and animals. Partially reduces fructose to mannitol.
Yeasts		Tolerate acid, 40% sugar aw > =0. 85	
Saccharomyces Cerevisiae S. pombe	Many aerobic, some anaerobes	pH 4-4. 5 20-30? C	S. cerevisiae can shift its metabolism from a fermentative to an oxidative pathway, depending on oxygen availability. Most yeasts produce alcohol and carbon dioxide from sugars.
Debaromyces Zygosaccharomyc			Tolerant of high salt

DESIRABLE FERMENTATION

It is essential with any fermentation to ensure that only the desired bacteria, yeasts or moulds start to multiply and grow on the substrate. This has the effect of suppressing other micro-organisms which may be either pathogenic and cause food poisoning or will generally spoil the fermentation process, resulting in an end-product which is neither expected or desired. An everyday example used to illustrate this point is the differences in spoilage between pasteurised and unpasteurised milk. Unpasteurised milk will spoil naturally to produce a sour tasting product which can be used in baking to improve the texture of certain breads.

Pasteurised milk, however, spoils (non-desirable fermentation) to produce an unpleasant product which has to be disposed of. The reason for this difference is that pasteurisation (despite being a very important process to destroy pathogenic micro-organisms) changes the microorganism environment and if pasteurised milk is kept untefrigerated for some time, undesirable micro-organisms start to grow and multiply before the desirable ones. In the case of unpasteurised milk, the non-pathogenic lactic acid bacteria start to grow and multiply at a greater rate that any pathogenic bacteria. Not only do the larger numbers of lactic acid bacteria compete more successfully for the available nutrients, but as they grow they produce lactic acid which increases the acidity of the substrate and further suppresses the bacteria which cannot tolerate an acid environment.

Most food spoilage Organisms cannot survive in either alcoholic or acidic environments. Therefore, the production of both these end products can prevent a food from spoilage and extend the shelf life. Alcoholic and acidic fermentations generally offer cost effective methods of preserving food for people in developing countries, where more sopHisticated means of preservation are unaffordable and therefore not an option.

The principles of microbial action are identical both in the use of micro-organisms in food preservation, such as through desirable fermentations, and also as agents of destruction via food spoilage. The type of organisms present and the environmental conditions will determine the nature of the reaction and the ultimate products. By manipulating the external reaction conditions, microbial reactions can be controlled to produce desirable results. There are several means of altering the reaction environment to encourage the growth of desirable Organisms.

MANIPULATION OF MICROBIAL GROWTH AND ACTIVITY

There are six major factors that influence the growth and activity of micro-organisms in foods. These are moisture, oxygen concentration, temperature, nutrients, pH and inhibitors. The food supply available to the micro-organisms depends on the composition of the food on which they grow. All micro-organisms differ in their ability to metabolise proteins, carbohydrates and fats. Obviously, by manipulating any of these six factors, the activity of micro-organisms within foods can be controlled.

Moisture

Water is essential for the growth and metabolism of all cells. If it is reduced or removed, cellular activity is decreased. For example, the removal of water from cells by drying or the change in state of water (from liquid to solid) affected by freezing, reduces the availability of water to cells (including microbial cells) for metabolic activity. The form in which water exists within the food is important as far as microbial activity is concerned. There are two types of water - free and bound. Bound water is present within the tissue and is vital to all the pHysiological processes within the cell.

Free water exists in and around the tissues and can be removed from cells without seriously interfering with the vital processes. Free water is essential for the survival and activity of micro-organisms. Therefore, by removing free water, the level of microbial activity can be controlled. The amount of water available for micro-organisms is referred to as the water activity (aw). Pure water has a water activity of 1.0.

Bacteria require more water than yeasts, which require more water than moulds to carry out their metabolic activities. Almost all microbial activity is inhibited below aw of 0.6. Most fungi are inhibited below aw of 0.7, most yeasts are inhibited below aw of 0.8 and most bacteria below aw 0.9. Naturally, there are exceptions to these guidelines and several species of micro-organism can exist outside the stated range. The water activity of foods can be changed by altering the amount of free water available.

There are several ways to achieve this - drying to remove water; freezing to change the state of water from liquid to solid; increasing or decreasing the concentration of solutes by adding salt or sugar or other hydropHylic compounds (salt and sugar are the two common additives used for food preservation). Addition of salt or sugar to a food will bind free water and so decrease the aw. Alternatively, decreasing the concentration will increase the amount of free water and in turn the aw. Manipulation of the aw in this manner can be used to encourage the growth of favourable micro-organisms and discourage the growth of spoilage ones.

Water Activity for Microbial Reactions

Aw	Phenomenon	Examples
1. 00		Highly perishable foods
0. 95	Pseudomonas, Bacillus, Clostridium perfringens and some yeasts inhibited	Foods with 40% sucrose or 7% salt
0. 90	Lower limit for bacterial growth. Salmonella, Vibrio parahaemolyticus, Clostridium botulinum, Lactobacillus and some yeasts and fungi inhibited	Foods with 55% sucrose, 12% salt. Intermediate-moisture foods (aw = 0. 90-0. 55)
0. 85	Many yeasts inhibited	Foods with 65% sucrose, 15% salt
0. 80	Lower limit for most enzyme activity and growth of most fungi. StapHylococcus aureus inhibited	Fruit syrups
0. 75	Lower limit for halopHilic bacteria	Fruit jams
0. 70	Lower limit for growth of most xeropHilic fungi	
0. 65	Maximum velocity of Maillard reactions	
0. 60	Lower limt for growth of osmopHilic or xeropHilic yeasts and fungi	Dried fruits (15- 20% water)
0. 55	Deoxyribose nucleic acid (DNA) becomes disordered (lower limit for life to continue)	
0. 50		Dried foods (aw=0-0. 55)
0. 40	Maximum oxidation velocity	
0. 25	Maximum heat resistance of bacterial spores	

OXIDATION-REDUCTION POTENTIAL

Oxygen is essential to carry out metabolic activities that support all forms of life. Free atmospHeric oxygen is utilised by some groups of micro-organisms, while others are able to metabolise the oxygen which is bound to other compounds such as carbohydrates. This bound oxygen is in a reduced form. Micro-organisms can be broadly classified into two groups - aerobic and anaerobic. Aerobes grow in the presence of atmospHeric oxygen while anaerobes grow in the absence of atmospHeric oxygen. In the middle of these two extremes are the facultative anaerobes which can adapt to the prevailing conditions and grow in either the absence or presence of atmospHeric oxygen.

MicroaeropHilic organisms grow in the presence of reduced amounts of atmospHeric oxygen. Thus, controlling the availability of free oxygen is one means of controlling microbial activity within a food. In aerobic fermentations, the amount of oxygen present is one of the limiting factors. It determines the type and amount of biological product obtained, the amount of substrate consumed and the energy released from the reaction. Moulds do not grow well in anaerobic conditions, therefore they are not important in terms of food spoilage or beneficial fermentation, in conditions of low oxygen availability.

Temperature

Temperature affects the growth and activity of all living cells. At high temperatures, organisms are destroyed, while at low temperatures, their rate of activity is decreased or suspended. Micro-organisms can be classified into three distinct categories according to their temperature preference.

Nutritional Requirements

The majority of organisms are dependent on nutrients for both energy and growth. Organisms vary in their specificity towards different substrates and usually only colonise foods which contain the substrates they require. Sources of energy vary from simple sugars to complex carbohydrates and proteins. The energy requirements of micro-organisms are very high. Limiting the amount of substrate available can check their growth.

Classification of Bacteria According to Temperature Requirements.

	Temperature Required for Growth °C			
Type of bacteria	**Minimum**	**Optimum**	**Maximum**	**General Sources of Bacteria**
PsychropHilic	0 to 5	15 to 20	30	Water and frozen foods
MesopHilic	10 to 25	30 to 40	35 to 50	Pathogenic and non-pathogenic bacteria
ThermopHilic	25 to 45	50 to 55	70 to 90	Spore forming bacteria from soil and water

Hydrogen Ion Concentration (pH)

The pH of a substrate is a measure of the hydrogen ion concentration. A food with a pH of 4. 6 or less is termed a high acid or acid food and will not permit the growth of bacterial spores. Foods with a pH above 4. 6. are termed low acid and will not inhibit the growth of bacterial spores. By acidifying foods and achieving a final pH of less than 4. 6, most foods are resistant to bacterial spoilage.

The optimum pH for most micro-organisms is near the neutral point (pH 7.0). Certain bacteria are acid tolerant and will survive at reduced pH levels. Notable acid-tolerant bacteria include the Lactobacillus and Streptococcus species, which play a role in the fermentation of dairy and vegetable products. Moulds and yeasts are usually acid tolerant and are therefore associated with spoilage of acidic foods.

Micro-organisms vary in their optimal pH requirements for growth. Most bacteria favour conditions with a near neutral pH (7). Yeasts can grow in a pH range of 4 to 4. 5 and moulds can grow from pH 2 to 8. 5, but favour an acid pH. The varied pH requirements of different groups of micro-organisms is used to good effect in fermented foods where successions of micro-organisms take over from each other as the pH of the environment changes. For instance, some groups of microorganisms ferment sugars so that the pH becomes too low for the survival of those microbes. The acidopHilic microorganisms then take over and continue the reaction. The affinity for different pH can also be used to good effect to occlude spoilage organisms.

Inhibitors

Many chemical compounds can inhibit the growth and activity of micro-organisms. They do so by preventing metabolism, denaturation of the protein or by causing pHysical damage to the cell. The production of substrates as part of the metabolic reaction also acts to inhibit microbial action.

CONTROLLED FERMENTATION

Controlled fermentations are used to produce a range of fermented foods, including sauerkraut, pickles, olives, vinegar, dairy and other products. Controlled fermentation is a form of food preservation since it generally results in a reduction of acidity of the food, thus preventing the growth of spoilage micro-organisms. The two most common acids produced are lactic acid and acetic acid, although small amounts of other acids such as propionic, fumaric and malic acid are also formed during fermentation. It is highly probable that the first controlled food fermentations came into existence through trial and error and a need to preserve foods for consumption later in the season. It is possible that the initial attempts at preservation involved the addition of salt or seawater. During the removal of the salt prior to consumption, the foods would pass through stages favourable to acid fermentation. Although the process worked, it is likely that the causative agents were unknown.

Today, there are numerous examples of controlled fermentation for the preservation and processing of foods. However, only a few of these have been studied in any detail - these include sauerkraut, pickles, kimchi, beer, wine and vinegar production. Although the general principles and processes for many of the fermented fruit and vegetable products are the same -relying mainly on lactic acid and acetic acid forming bacteria, yeasts and moulds, the reactions have not been quantified for each product. The reactions are usually very complex and involve a series of micro-organisms, either working together or in succession to achieve the final product.

YEAST FERMENTATIONS

WHAT ARE YEASTS?

A yeast is a unicellular fungus which reproduces asexually by budding or division, especially the genus Saccharomyces which is important in food fermentations. Yeasts and yeast like fungi are widely distributed in nature. They are present in orchards and vineyards, in the air, the soil and the intestinal tract of animals. Like bacteria and moulds, they can have beneficial and non-beneficial effects in foods. Most yeasts are larger than most bacteria. The most well known examples of yeast fermentation are in the production of alcoholic drinks and the leavening of bread. For their participation in these two processes, yeasts are of major importance in the food industry. Some yeasts are chromogenic and produce a variety of pigments, including green, yellow and black. Others are capable of synthesising essential B group vitamins.

Although there is a large diversity of yeasts and yeast like fungi, (about 500 species), only a few are commonly associated with the production of fermented foods. They are all either ascomycetous yeasts or members of the genus Candida. Varieties of the Saccharomyces cervisiae genus are the most common yeasts in fermented foods and beverages based On fruit and vegetables. All strains of this genus ferment glucose and many ferment other plant derived carbohydrates such as sucrose, maltose and raffinose. In the tropics, Saccharomyces pombe is the dominant yeast in the production of traditional fermented beverages, especially those derived from maize and millet

CONDITIONS NECESSARY FOR FERMENTATION

Most yeasts require an abundance of oxygen for growth, therefore by controlling the supply of oxygen, their growth can be checked. In addition to oxygen, they require a basic substrate such as sugar. Some yeasts can ferment sugars to alcohol and carbon dioxide in the absence of air but require oxygen for growth. They produce ethyl alcohol and carbon dioxide from simple sugars such as glucose and fructose.

$$C_6H_{19}O_7 \rightarrow 2C_2H_5OH + 2CO_2$$

In conditions of excess oxygen (and in the presence of acetobacter) the alcohol can be oxidised to form acetic acid. This is undesirable if the end product is a fruit alcohol, but is a technique employed for the production of fruit vinegars.

Yeasts are active in a very broad temperature range - from 0 to 50 C, with an optimum temperature range of 20 to 30 C. The optimum pH for most micro-organisms is near the neutral point (pH 7.0). Moulds and yeasts are usually acid tolerant and are therefore associated with the spoilage of acidic foods. Yeasts can grow in a pH range of 4 to 4.5 and moulds can grow from pH 2 to 8.5, but favour an acid pH.

In terms of water requirements, yeasts are intermediate between bacteria and moulds. Bacteria have the highest demands for water, while moulds have the least need. Normal yeasts require a minimum water activity of 0.85 or a relative humidity of 88%.

Yeasts are fairly tolerant of high concentrations of sugar and grow well in solutions containing 40% sugar. At concentrations higher than this, only a certain group of yeasts the osmopHilic type can survive. There are only a few yeasts that can tolerate sugar concentrations of 65-70% and these grow very slowly in these conditions. Some yeasts for example the Debaromyces can tolerate high salt concentrations. Another group which can tolerate high salt concentrations and low water activity is Zygosaccharomyces rouxii, which is associated with fermentations in which salting is an integral part of the process.

PRODUCTION OF FRUIT ALCOHOL

Alcohol and acids are two primary products of fermentation, both used to good effect in the preservation of foods. Several alcohol-fermented foods are preceded by an acid fermentation and in the presence of oxygen and acetobacter, alcohol can be fermented to produce acetic acid. Most food spoilage organisms cannot survive in either alcoholic or acidic environments. Therefore, the production of both these end products can prevent a food from undergoing spoilage arid extend its shelf life.

Primitive wines and beers have been produced, with the aid of yeasts, for thousands of years, although it was not until about four hundred years ago that micro-organisms associated with the fermentation were observed and identified. It was not until the 1850's that Louis Pasteur demonstrated unequivocally the involvement of yeasts in the production of wines and beers. Since then, the knowledge of yeasts and the conditions necessary for fermentation of wine and beer has increased to the point where pure culture fermentations are now used to ensure consistent product quality. Originally, alcoholic fermentations would have been spontaneous events that resulted from the activity of micro-organisms naturally present. These non-scientific methods are still used today for the home preparation of many of the worlds traditional beers and wines.

Alcoholic drinks fall into two broad categories: wines and beers. Wines are made from the juice of fruits and beers from cereal grains. The principal carbohydrates in fruit juices are soluble sugars; the principal carbohydrate in grains is starch, an insoluble polysaccharide. The yeasts that bring about alcoholic fermentation can attack soluble sugars but do not produce starch-splitting enzymes. Wines can therefore be made by the direct fermentation of the raw material, while the production of beer requires the hydrolysis of starch to yield sugars fermentable by yeast, as a preliminary step.

Raw fruit juice is usually a strongly acidic solution, containing from 10 to 25 percent soluble sugars. Its acidity and high sugar concentration make it an unfavourable medium for the growth of bacteria but highly suitable for yeasts and moulds. Raw fruit juice naturally contains many yeasts, moulds, and bacteria, derived from the surface of the fruit. Normally the yeast used in alcoholic fermentation is a strain of the species Saccharomyces cerevisiae. The fermentation may be allowed to proceed spontaneously, or can be "started" by inoculation with a must that has been previously successfully fermented by S. cerevisiae var. ellipsoideus. Many modern wineries eliminate the original microbial population of the must by pasteurisation or by treatment with sulpHur dioxide. The must is then inoculated with a starter culture derived from a pure culture of a suitable strain of wine yeast.

This procedure eliminates many of the uncertainties and difficulties of older methods. At the start of the fermentation, the must is aerated slightly to build up a large and vigorous yeast population; once fermentation sets in, the rapid production of carbon dioxide maintains anaerobic conditions, which prevent the growth of undesirable aerobic organisms, such as bacteria and moulds. The temperature of fermentation is usually from 25 to 30°C, and the duration of the fermentation process may extend from a few days to two weeks. As soon as the desired degree of sugar disappearance and alcohol production has been attained, the microbiological pHase of wine making is over. Thereafter, the quality and stability of the wine depend very largely on preventing further microbial activity, both during the "aging" in wooden casks and after bottling.

At all stages during its manufacture, fruit juice alcohol is subject to spoilage by undesirable microorganisms. Pasteur, whose descriptions of the organisms responsible and recommendations for overcoming them are still valid today, first scientifically explored the problem of the "diseases" of wines. The most serious aerobic spoilage processes are brought about by film-forming yeasts and acetic acid bacteria, both of which grow at the expense of the alcohol, converting it to acetic acid or to carbon dioxide and water. The chief danger from these organisms arises when access of air is not carefully regulated during aging. Much more serious are the diseases caused by fermentative bacteria, particularly rod-shaped lactic acid bacteria, which utilise any residual sugar and impart a mousy taste to the wine. Such wines are known as turned wines. Since oxygen is unnecessary for the growth of lactic acid bacteria, wine spoilage of this kind can occur even after bottling. These risks of spoilage can be minimised by pasteurisation after bottling.

GRAPE WINE

Grape wine is perhaps the most common fruit juice alcohol. Because of the commercialisation of the product for industry, the process has received most research attention and is documented in detail. The production of grape wine involves the following basic steps: crushing the grapes to extract the juice; alcoholic fermentation; maltolactic fermentation if desired; bulk storage and maturation of the wine in a cellar; clarification and packaging. Although the process is fairly simple, quality control demands that the fermentation is carried out under controlled conditions to ensure a high quality product. The distinctive flavour of grape wine originates from the grapes as raw material and subsequent processing operations. The grapes contribute trace elements of many volatile substances (mainly terpenes) which give the final product the distinctive fruity character. In addition, they contribute non-volatile compounds (tartaric and malic acids) which impact on flavour and tannins which give bitterness and astringency.

The latter are more prominent in red wines as the tannin components are located in the grape skins. Although yeasts are the principal organisms involved, filamentous fungi, lactic acid bacteria, acetic acid bacteria and other bacterial groups all play a role in the production of alcoholic fruit products. Normal grapes harbour a diverse micro-flora, of which the principal yeasts (Saccharomyces cerevisiae) involved in desirable fermentation are in the minority. Lactic acid bacteria and acetic acid bacteria are also present. The proportions of each and total numbers present are dependent upon a number of external environmental factors including the temperature, humidity, stage of maturity, damage at harvest and application of fungicides. It is essential to ensure proliferation of the desired species at the expense of the non-desired ones. This is achieved through ensuring fermentation conditions are such to encourage Saccharomyces species. The fermentation may be initiated using a starter culture of Saccharomyces cerevisiae in which case the juice is inoculated with populations of yeast of 106 to 107 cfu/ml juice. This approach produces a wine of generally expected taste and quality.

If the fermentation is allowed to proceed naturally, utilising the yeasts present on the surface of the fruits, the end result is less controllable, but produces wines with a range of flavour characteristics. It is likely that natural fermentations are practiced widely around the world, especially for home production of wine. During alcoholic fermentation, yeasts are the prominent species.

The composition of fruit juice - its acid and sugar level and low pH favour the growth of yeasts and production of ethanol that restricts the growth of bacteria and fungi. In natural fermentations, there is a progressive pattern of yeast growth. Several species of yeast, including Kloeckera, Hanseniaspora, Candida and Metschnikowia, are active for the first two to three days of fermentation. The build up of end products (ethanol) is toxic to these yeasts and they die off, leaving Saccharomyces cerevisiae to continue the fermentation to the end. S. cerevisiae can tolerate much higher levels of ethanol (up to 15% v/v or more) than the other species who only tolerate upto 5 or 8% alcohol. Because of its tolerance of alcohol, S. cerevisiae dominates wine fermentation and is the species that has been commercialised for starter cultures.

Traditionally, fermentation was carried out in large wooden barrels or concrete tanks. Modern wineries now use stainless steel tanks as these are more hygienic and provide better temperature control. White wines are fermented at 10 to 18° C for about seven to fourteen days. The low temperature and slow fermentation favours the retention of volatile compounds. Red wines are fermented at 20 to 30°C for about seven days. This higher temperature is necessary to extract the pigment from the grape skins.

FACTORS AFFECTING WINE FERMENTATION

There are several variables which can affect the fermentation process and final quality of wine. Factors which are most important to control are:

- The clarification and pre-treatment of juice
- Chemical composition of the juice
- Temperature of the fermentation
- The influences of other micro-organisms.

Clarification and pre-Treatment of Juice

Excessive clarification removes many of the natural yeasts and flora. This is beneficial if a tightly controlled induced fermentation is desired, but less so if the fermentation is a natural one. Long periods of settling out however, encourage the growth of natural flora, which can contribute to the fermentation.

Chemical Composition of Juice

The main consituents of grape juice are glucose (75 to 150 g/1), fructose (75 to 150 g/1), tartaric acid (2 to 10 g/1), malic acid (1 to 8 g/1) and free amino acids (0.2 to 2.5 g/1). The main reaction is the fermentation of glucose and fructose to ethanol and carbon dioxide. However, the presence of nitrogenous and sulpHurous products also contributes to the fermentation. The addition of sulpHur dioxide to the juice delays the growth of yeast, but does not necessarily inhibit growth of the non- Saccharomyces strains. Fruits generally contain sufficient substrates soluble sugars for the yeast to ferment and convert into an acceptable concentration of alcohol. Sugar can be added to fruit juices with a low sugar content, to increase the amount of fermentable substrate.

Temperature

Temperature has ah impact on the growth and activity of different strains of yeast. At temperatures of 10 to 15 C, the non-Saccharomyces species have an increased tolerance to alcohol and therefore have the potential to contribute to the fermentation.

Influence of other Micro-Organisms

Other micro-organisms have the potential to influence wine production at all stages of the process. Prior to harvest, yeasts grow on the surface of grapes. Fungicides are used in an attempt to control their growth, but these disturb the natural balance of flora, thus making it difficult to carry out a 'natural' fermentation.

Overuse of fungicides can lead to the development of resistant strains of yeast which have the potential to produce toxins which destroy the desirable yeast species. These yeasts are known as 'killer' strains. Other microbes have further chances to influence the fermentation during the clarification process, after fermentation and during maturation and bottling when acetobacter species can oxidise the alcohol and produce acetic acid.

About two to three weeks after the alcoholic fermentation is finished wines often undergo a malo-lactic fermentation. This occurs naturally and lasts for about four weeks. It is a lactic acid fermentation, initiated by lactic acid bacteria resident in the wine. Inoculating the fermented wine with cultures of Leuconostoc oenos can start the process if it is desired. The main reaction of these bacteria is the decarboxylation of L-malic acid to L-lactic acid, which decreases the acidity of the wine and increases its pH by about 0.3 to 0.5 units.

Wines produced from grapes grown in colder climates tend to have a higher concentration of malic acid and a lower pH (3.0 to 3.5) and the taste benefits from this slight decrease in acidity. The benefits of this process are that it imparts a more mellow flavour to the wine. The growth of malo-lactic bacteria also contributes to the taste of the wine. Wines that have undergone a malo-lactic fermentation appear to be less susceptible to any further damage from other bacteria. This could be because L. oenos has used up all available substrate, or it may have secreted bacteriocins which prevent the growth of other species.

Although the malo-lactic fermentation seems to be a useful process, not all wines benefit from it. Wines produced from grapes in warmer climates tend to be less acidic (pH > 3.5) and a further reduction in acidity may have adverse effects on the quality of the wine. Decreasing the acidity also increases the pH to values which can allow spoilage organisms to multiply. It is difficult to prevent the malo-lactic fermentation from taking place naturally, especially later on after the wine has been bottled.

In low acid wines, the acidity may be adjusted after this fermentation has taken place. The malo-lactic fermentation can be prevented by controlling several factors: the wine pH (<3.2); ethanol content (> 14%) and levels of sulpHur dioxide (>50mg/l). The bacteriocin nisin can also be used to control the growth of malo-lactic bacteria. However the subtle blend of aromas and flavours that contribute to the final taste may be lost by such stringent control.

The conversion of malic acid to lactic acid is one of the main reactions carried out by wine lactic acid bacteria. L oenos needs to be present in significant numbers (greater than 106 cfu/ml) for the reaction to take place at a suitable pace. The bacteria use residual pentose and hexose sugars in the wine as a substrate for growth. The main reaction is the deacidification (or decarboxylation) of malic acid. In addition to this, the by-products of the reaction impart flavours and aromas to the wine.

During storage, wines are prone to non-desirable microbial changes. Yeasts, lactic acid bacteria, acetic acid bacteria and fungi can all spoil or taint wines after the fermentation process is completed. The changes that occur are increased acidification through the formation of acetic and other acids from alcohol; increased carbonation through a secondary fermentation of residual sugars and flavour changes through the metabolism of numerous compounds.

PRODUCTS OF YEAST FERMENTATATION

The major products of yeast fermentation are alcoholic drinks and bread. With respect to fruits and vegetables, the most important products are fermented fruit juices and fermented plant saps. Virtually any fruit or sugary plant sap can be processed into an alcoholic beverage.

The process is well known being essentially an alcoholic fermentation of sugars to yield alcohol and carbon dioxide. It should be noted that alcohol production requires special licences or is prohibited in many countries.

FERMENTED FRUIT JUICES

There are many fermented drinks made from fruit in Africa, Asia and Latin America. These include drinks made from bananas, grapes and other fruit. Grape wine is perhaps the most economically important fruit juice alcohol. It is of major economic importance in Chile, Argentina, South Africa, Georgia, Morocco and Algeria.

Because of the commercialisation of the product for industry, the process has received most research attention and is documented in detail. Banana beer is probably the most wide spread alcoholic fruit drink in Africa and is of cultural importance in certain areas. Alcoholic fruit drinks are made from many other fruits including dates in North Africa, pineapples in Latin America and jack fruits in Asia.

RED GRAPE WINE

Product Description

Red grape wine is an alcoholic fruit drink of between 10 and 14% alcoholic strength. The colour ranges from a light red to a deep dark red. It is made from the fruit of the grape plant (Vitis vinifera). There are many varieties of grape used including Cabernet Sauvignon, Grenache, Nebbiolo, Pinot Noir, and Torrontes. The skins of the grape are allowed to be fermented in red wine production, to allow for the extraction of colour and tannins, which contribute to the flavour. The grapes contribute trace elements of many volatile substances, which give the final product the distinctive fruity character. In addition, they contribute non-volatile compounds (tartaric and malic acids) which impact on flavour and tannins, which give bitterness and astringency.

Raw Material Preparation

Ripe and undamaged grapes should be used. Red grapes are crushed to yield the juice plus skins, which is known as must.

Processing

The crushed grapes are transferred to fermentation vessels. The ethanol formed during this fermentation assists with the extraction of pigments from the skins. This takes between 24 hours and three weeks depending on the colour of the final product required.

The skins are then removed and the partially fermented wine is transferred to a separate tank to complete the fermentation. The fermentation can be from naturally occurring yeasts on the skin of the grape or using a starter culture of Saccharomyces cerevisiae in which case the juice is inoculated with populations of yeast.

This approach produces a wine of generally expected taste and quality. If the fermentation is allowed to proceed naturally, utilising the yeasts present on the surface of the fruits, the end result is less controllable, but produces wines with a range of flavour characteristics.

Traditionally, fermentation was carried out in large wooden barrels or concrete tanks. Modem wineries now use stainless steel tanks as these are more hygienic and provide better temperature control. Fermentation stops naturally when all the fermentable sugars have been converted to alcohol or when the alcoholic strength reaches the limit of tolerance of the strain of yeast involved.

Fermentation can be stopped artificially by adding alcohol, by sterile filtration or centrifugation. Some wines can be drunk immediately. However most wines develop distinctive favours and aromas by ageing in wooden casks.

Packaging and Storage

Traditionally wine was delivered to the point of sale in casks. The product is traditionally packaged in glass bottles with corks, made from the bark of the cork oak (Quercus suber). The bottles should be kept out of direct sunlight. During storage, wines are prone to non-desirable microbial changes. Yeasts, lactic acid bacteria, acetic acid bacteria and fungi can all spoil or taint wines after the fermentation process is completed.

Flow Diagram

Selection of grapes ↓	Mature and Undamaged grapes
Crushing ↓	Traditionally manually, but now by crushers
Pre-fermentation ↓	24 hours to three weeks depending on colour required
Removal of skin ↓	Can add sulpHur dioxide to inhibit wild yeasts
Fermentation ↓	
Maturation ↓	Ageing to develop aromas and flavours

WHITE GRAPE WINE

Product Description

White grape wine is an alcoholic fruit drink of between 10 and 14% alcoholic strength. It is prepared from the fruit of the grape plant (Vitis vinifera), and is pale yellow in colour. There are many varieties used including Airen, Chardonnay, Palomino, Sauvignon Blanc and Ugni Blanc. The main difference between red and white wines is the early removal of grape skins in white wine production. The distinctive flavour of grape wine originates from the grapes as raw material and subsequent processing operations. The grapes contribute trace elements of many volatile substances (mainly terpenes) which give the final product the distinctive fruity character.

Preparation of Raw Materials

Ripe and undamaged grapes should be used. The grapes are crushed to yield the juice and the skins are removed and separated out Sometimes the juice is clarified by allowing it to stand for 24 to 48 hours at 5 to 10 C, by filtering or centrifugation. Pectolytic enzymes may be added to accelerate the breakdown of cell wall tissue and to improve the clarity of juice. Excessive clarification removes many of the natural yeasts and flora. This is beneficial if a tightly controlled induced fermentation is desired, but less so if the fermentation is a natural one. Long periods of settling out however, encourage the growth of natural flora, which can contribute to the fermentation.

Processing

The clarified juice is transferred to a fermentation tank where fermentation either begins spontaneously or is induced by the addition of a starter culture. Traditionally, fermentation was carried out in large wooden barrels or concrete tanks. Modern wineries now use stainless steel tanks as these are more hygienic and provide better temperature control. White wines are fermented at 10 to 18° C for about seven to fourteen days. The low temperature and slow fermentation favours the retention of volatile compounds.

The fermentation can be from naturally occurring yeasts on the skin of the grape or using a starter culture of Saccharomyces cerevisiae. This approach produces a wine of generally expected taste and quality. If the fermentation is allowed to proceed naturally, utilising the yeasts present on the surface of the fruits, the end result is less controllable, but produces wines having a range of flavour characteristics.

It is likely that natural fermentations are practised widely around the world, especially for home production of wine. During storage, wines are prone to non-desirable microbial changes. Yeasts, lactic acid bacteria, acetic acid bacteria and fungi can all spoil or taint wines after the fermentation process is completed.

Packaging and Storage

Traditionally wine was delivered to the point of sale in casks. The product is traditionally packaged in glass bottles with corks, made from the bark of the cork oak (Quercus suber). The bottles should be kept out of direct sunlight. During storage, wines are prone to non-desirable microbial changes. Yeasts, lactic acid bacteria, acetic acid bacteria and fungi can all spoil or taint wines after the fermentation process is completed.

Flow Diagram

Selection of grapes ↓	Mature and undamaged grapes
Crushing ↓	Traditionally manual but now usually by crushers
Removal of skins ↓	
Clarification ↓	By standing, filtration or centrifugation
Fermentation ↓	
Ageing ↓	

BANANA BEER

Product Description

Banana beer is made from bananas, mixed with a cereal flour (often sorghum flour) and fermented to an orange, alcoholic beverage. It is sweet and slightly hazy with a shelf-life of several days under correct storage conditions. There are many variations in how the beer is made. For instance Urwaga banana beer in Kenya is made from bananas and sorghum or millet and Lubisi is made from bananas and sorghum.

Preparation of Raw Materials

Ripe bananas Musa spp. are selected. The bananas should be peeled. If the peels cannot be removed by hand then the bananas are not sufficiently ripe.

Processing

The first step of the process is the extraction of banana juice. Extraction of a high yield of banana juice without excessive browning or contamination by spoilage microorganisms and proper filtration to produce a clear product is of great importance. Grass is used as an aid in obtaining clarified juice.

One volume of water is added to every three volumes of banana juice. This makes the total soluble solids low enough for the yeast to act. Cereals are ground and roasted and added to improve the colour and flavour of the final product. The mixture is placed in a container, which is covered in polythene to ferment for 18 to 24 hours. The raw materials are not sterilised by boiling and therefore provide an excellent substrate for microbial growth.

Flow Diagram

Raw materials ↓	Ripe bananas
Peel ↓	Peel by hand
Remove residue ↓	Use grass to knead or squeeze out the juice
Mix with water ↓	The water: banana juice ratio should be 1: 3
Mix with cereals ↓	Mix with ground and roasted cereals to local taste
Ferment ↓	In plastic container. Leave to ferment for 18 to 24 hours.
Filter ↓	Through cotton cloth
Pack ↓	Store

It is essential that proper hygienic procedures are followed and that all equipment is thoroughly sterilised to prevent contaminating bacteria from competing with the yeast and producing acid instead of alcohol. This can be done by cleaning with boiling water or with chlorine solution. Care is necessary to wash the equipment free of residual chlorine as this would interfere with the actions of the yeast. Strict personal hygiene is also essential.

For many traditional fermented products, the microorganisms responsible for the fermentation are unknown to scientists. However there has been research to identify the micro-organisms involved in banana beer production. The main micro-organism involved, is Saccharomyces cerevisiae which is the same organism involved in the production of grape wine. However many other micro-organisms associated with the fermentation have been identified. These varied according to the region of production. After fermentation the product is filtered through cotton cloth.

Packaging and Storage

Packaging is usually only required to keep the product for its relatively short shelf-life. Clean glass or plastic bottles are used. The product is kept in a cool place away from direct sunlight.

CASHEW WINE

Product Description

Cashew wine is a light yellow alcoholic drink prepared from the fruit of the cashew tree (Ancardium occidentale). It contains an alcohol content of between 6 and 12% alcohol.

Preparation of Raw Materials

In gathering the fruits and transporting them to the workshop, the prime purpose should be to have the fruit arrive in the very best condition possible. Cashew apples are sorted and only mature undamaged cashew apples should be selected. These should be washed in clean water.

Processing

The cashew apples are cut into slices to ensure a rapid rate of juice extraction when crushed in a juice press. The fruit juice is sterilised in stainless steel pans at a temperature of 85°C in order to eliminate wild yeast. The juice is filtered and treated either sodium or potassium metabisulpHite to destroy or inhibit the growth of any undesirable types of micro-organisms - acetic acid bacteria, wild yeasts and moulds.

Flow Diagram

Selection ↓	Mature, sound cashew apples
Slicing ↓	To increase extraction of juice
Crushing ↓	
Sterilisation ↓	At 850 C
Filtered ↓	
Inoculation ↓	With *S. cerevisiae*
Fermentation ↓	Two weeks
Filtered ↓	
Pasteurisation ↓	50-60? C
Ageing ↓	In wooden vats for 6 months

Wine yeast (Saccharomyees cerevisiae - var ellipsoideus) are added. Once the yeast is added, the contents are stirred well and allowed to ferment for about two weeks. The wine is separated from the sediment. It is clarified by using fining agents such as gelatin, pectin or casein which are mixed with the wine. Filtration is carried out with filter-aids such as fullers earth. The filtered wine is transferred to wooden vats. The wine is then pasteurised at 50°-60°C. Temperature should be controlled, so as not to heat it to about 70°C, since its alcohol content would vaporise at a temperature of 75°-78°C. It is then stored in wooden vats and subjected to ageing. At least six months should be allowed for ageing. If necessary, wine is again clarified prior to bottling. During ageing, and subsequent maturing in bottles many reactions, including oxidation, occur with the formation Of traces of esters and aldehydes., which together with the tannin and acids already present enhance the taste, aroma and preservative properties of the wine.

Packaging and Storage

The product is packaged in glass bottles with corks. The bottles should be kept out of direct sunlight.

Tepache

Tepache is a light, refreshing beverage prepared and consumed throughout Mexico. In the past, tepache was prepared from maize, but nowadays various fruits such as pineapple, apple and orange are used. The pulp and juice of the fruit are allowed to ferment for one or two days in water with some added brown sugar. The mixture is contained in a lidless wooden barrel called a 'tepachera', which is covered with cheese cloth.

After a day or two, the tepache is a sweet and refreshing beverage. If fermentation is allowed to proceed longer, it turns into an alcoholic beverage and later into vinegar. The microorganisms associated with the product include Bacillus subtilis, B. graveolus and the yeasts, Torulopsis insconspicna, Saccharomyces cerevisiae and Candida queretana.

Colonche

Colonche is a sweet, fizzy beverage produced in Mexico by fermenting the juice of the fruits of the prickly pear cacti - mainly Opuntia species. The procedure for preparing colonche is essentially the same as has been followed for centuries. The cactus fruits are peeled and crushed to obtain the juice, which is boiled for 2-3 hours. After cooling, the juice is allowed to ferment for a few days. Sometimes old colonche or tibicos may be added as a starter. Tibicos are gelatinous masses of yeasts and bacteria, grown in water with brown sugar. They are also used in the preparation of tepache.

FORTIFIED GRAPE WINES

Fortified wines are made in the Republic of South Africa and North Africa. Fortified wine's are made by adding spirits to wines, either during or after fermentation, with the result that the alcohol content of the wines is raised to around 20 percent, i.e. approximately double that of table wines.

DATE WINE

Date wines are popular in Sudan and North Africa. They are made using a variety of methodologies. Dakhai is produced by placing dates in a clean earthenware pot. For every one volume of dates between two and four volumes of boiling water are added. This is allowed to cool and is then sealed for three days. More warm water is then added and the container sealed again for seven to ten days.

Many variations of date wine exist: El madfuna is produced by burying the earthenware pots underground. Benti merse is produced from a mixture of sorghum and dates. Nebit is produced from date syrup.

SPARKLING GRAPE WINE

Sparkling grape wines are made in the Republic of South Africa. Sparkling wines can be made in one of three ways. The cheapest method is to carbonate wines under pressure. Unfortunately, the sparkle of these wines quickly disappears, and the product is considered inferior to the sparkling wines produced by the traditional method of secondary fermentation.

This involves adding a special strain of wine yeast - a champagne yeast - to wine that has been artificially sweetened. Carbon dioxide produced by fermentation of the added sugar gives the wine its sparkle. In the original champagne method, which is still widely used today, this secondary fermentation is carried out in strong bottles, capable of withstanding pressure but early in the nineteenth century a method of fermenting the wine in closed tanks was devised, this being considerably cheaper than using bottles.

JACK-FRUIT WINE

Jack-fruit wine is an alcoholic beverage made by ethnic groups in the eastern hilly areas of India. As its name suggests, it is produced from the pulp of jack-fruit (Artocarpus heteropHyllus). Ripe fruit is peeled and the skin discarded. The seeds are removed and the pulp soaked in water. Using bamboo baskets, the pulp is ground to extract the juice, which is collected in earthenware pots. A little water is added to the pots along with fermented wine inoculum from a previous fermentation. The pots are covered with banana leaves and allowed to ferment at 18 to 30°C for about one week. The liquid is then decanted and drunk. During fermentation, the pH of the wine reaches a value of 3.5 to 3.8, suggesting that an acidic fermentation takes place at the same time as the alcoholic fermentation. Final alcohol content is about 7 to 8% within a fortnight.

FERMENTED PLANT SAPS

Virtually any sugary plant sap can be processed into an alcoholic beverage. The process is well known being essentially an alcoholic fermentation of sugars to yield alcohol and carbon dioxide. Many alcoholic drinks are made from the juices of plants including coconut palm, oil palm, wild date palm, nipa palm, rapHia palm and kithul palm.

PALM WINE

Product Description

Palm wine can be consumed in a variety of flavours varying from sweet unfermented to sour fermented and vinegary alcoholic drinks. There are many variations and names including emu and ogogoro in Nigeria and nsafufuo in Ghana. It is produced from sugary palm saps. The most frequently tapped palms are rapHia palms (RapHia hookeri or R. vinifera) and the oil palm (Elaeis guineense). Palm wine has been found to be nutritious.

The fermentation process increases the levels of thiamin, riboflavin, pyridoxin and vitamin B12. Like many African alcoholic beverages, palm wine has a very short shelf-life. The product is not preserved for more than one day. After this time accumulation of an excessive amount of acetic acid makes it unacceptable to consumers. The bark of a tree (Saccoglottis gabonensis) may be added as a preservative. The alkaloid and pHenolic compounds which are extracted into the wine have antimicrobial effect.

Preparation of Raw Materials

Sap is collected by tapping the palm. Tapping is achieved by making an incision between the kernels and a gourd is tied around to collect the sap which is collected a day or two later. The fresh palm juice is a sweet, clear, colourless juice containing 10-12 percent sugar and is neutral. The quality of the final wines is determined mostly by the conditions used in the collection of the sap. Often the collecting gourd is not washed between collections and residual yeasts in the gourd quickly begin the fermentation.

Processing

The sap is not heated and the wine is an excellent substrate for microbial growth. It is therefore essential that proper hygienic collection procedures are followed to prevent contaminating bacteria from competing with the yeast and producing acid instead of alcohol.

Fermentation starts soon after the sap is collected and within an hour or two, the sap becomes reasonably high in alcohol (up to 4%). If allowed to continue to ferment for more than a day, the sap begins turning into vinegar, although the vinegary flavour is preferred by some. Organisms responsible include S. cerevisiae, and Schizosaccharomyces pombe, and the bacteria Lactobacillus plantarum and L. mesenteroides.

There are reports that the yeasts and bacteria originate from the gourd, palm tree, and tapping implements. However the high sugar content of the juice would seem to selectively favour the growth of yeasts which might originate from the air.

This is supported by the fact that fermentation also takes place in plastic containers. Within 24 hours the initial pH is reduced from 7.4-6.8 to 5.5 and the alcohol content ranges from 1.5 to 2.1 percent. Within 72 hours the alcohol levels increase from 4.5 to 5.2 percent and the pH is 4.0. Organic acids present are lactic acid, acetic acid and tartaric acid.

The main control points are extraction of a high yield of palm sap without excessive contamination by spoilage microorganisms, and proper storage to allow natural fermentation to take place.

Flow diagram

Cut ↓	Cut 10-15 cm from the top of the trunk
Tapping ↓	A gourd is fixed below the cut
Collection ↓	The sap is collected each day
Fermentation ↓	Natural fermentation starts as soon as sap is collected
Filter ↓	Optional
Bottling ↓	Clean bottles should be used

Packaging and Storage

Packaging is usually only required to keep the product for its relatively short shelf-life. Clean glass or plastic bottles should be used. The product should be kept in a cool place away from direct sunlight.

TODDY

Product Description

Toddy is an alcoholic drink made by the fermentation of the sap from a coconut palm. It is white and sweet with a characteristic flavour. It is between 4 and 6% alcohol and has a shelf life of about 24 hours.

PREPARATION OF RAW MATERIALS

The sap is collected by slicing off the tip of an unopened flower. The sap oozes out and can be collected in a small pot tied underneath the flower.

Processing

The fermentation starts as soon as the sap collects in the pots on the palms, particularly if a small amount of toddy is left in the pots. The toddy is fully fermented in six to eight hours. The product is usually sold immediately due to its short shelf-life.

Flow Diagram

Tapping ↓	A small pot is fixed below the cut
Collection ↓	The sap is collected each day
Fermentation ↓	Natural fermentation starts as soon as sap is collected
Packaging ↓	Clean bottles should be used

Packaging and Storage

Packaging is usually only required to keep the product for its relatively short shelf-life. This is usually clean glass or plastic bottles. The product should be kept in a cool place away from direct sunlight.

Product Description

Pulque is a milky, slightly foamy, acidic and somewhat viscous beverage. It is obtained by fermentation of aguamiel, which is the name given to the juices of various cacti, notably Agave atrovirens and A. americana which are often called the "Century plant" in English. The alcohol content on pulque varies between six and seven percent.

The beverage obtained upon distilling pulque is called "Mezcal", and if manufactured in the Tequila region from a numbered distillery, it is referred to as "Tequila". The drink is often considered an apHrodisiac. The name Ticyaol is given to a good strain that makes one particularly virile. Pulque is frequently the potion of choice used by women during menstruation.

Preparation of Raw Materials

The juices are extracted from the plants when they are eight to ten years old and fermentation takes place spontaneously, although occasionally the juices are inoculated with a starter from previous fermentations.

Processing

The juice is allowed to ferment naturally through a mixed fermentation although yeast (Saccharomyces carbajali) is the main actor. Lacto-bacillus plantarum produces lactic acid and the viscosity of pulque is caused by the activity of two species of Leuconostoc which produce dextrans.

During fermentation of the juices of the plant, the soluble solids are reduced from between 25-30% to 6%; the pH falls from 7.4 to between 3.5 and 4.0; the sucrose content falls from 15% to 1% and vitamin levels are increased. For instance the vitamin content (milligrams of vitamins per 100g of product) increases from 5 to 29 for thiamine, 54 to 515 for niacin and 18 to 33 for riboflavin.

Packaging and Storage

Packaging is only required to keep the product for its relatively short shelf-life. Clean glass or plastic bottles should be used. The product should be kept in a cool place away from direct Sunlight.

ULANZI (BAMBOO WINE)

Product Description

Ulanzi is a fermented bamboo sap obtained by tapping young bamboo shoots during the rainy season. It is a clear, whitish drink with a sweet and alcoholic flavour.

Preparation of Raw Materials

The bamboo shoots should be young in order to obtain a high yield of sap. The growing tip is removed and a container fixed in place to collect the sap. The container should be clean in order to prevent contamination of the fresh sap.

Processing

The raw material is an excellent substrate for microbial growth and fermentation begins immediately after collection. Fermentation takes between five and twelve hours depending on the strength of the final product desired.

BASI (SUGAR CANE WINE)

Basi is a sugar cane wine made in the Philipppines by fermenting boiled, freshly extracted, sugar cane juice. A dried powdered starter is used to initiate the fermentation. The mixture is allowed to ferment for up to three months, and to age for up to one year. The final product is light brown in colour and has a sweet and a sour flavour. A similar product called shoto sake is made in Japan.

Muratina

Muratina is an alcoholic drink made from sugar cane and muratina fruit in Kenya. The fruit is cut in half, sun dried and boiled in water. The water is removed and the fruit is again sun dried. The fruit is added to a small amount of sugar cane juice and incubated in a warm place for 24 hours, after which it is removed and sun dried. The dried fruit is then added to a barrel of sugar cane juice which is allowed to ferment for between one and four days. The final product has a sour alcoholic taste.

BACTERIAL FERMENTATIONS

WHAT ARE BACTERIA

Bacteria are "a large group of unicellular or multi-cellular organisms lacking chloropHyll, with a simple nucleus, multiplying rapidly by simple fission, some species developing a highly resistant resting (spore) pHase; some species reproduce sexually, and some are motile. In shape they are spHerical, rodlike, spiral, or filamentous. They occur in air, water, soil, rotting organic material, animals and plants. SapropHytic forms are more numerous than parasites. A few forms are autotropHic".

There are several bacterial families present in foods, the majority of which are concerned with food spoilage. The important role of bacteria in the fermentation of foods is often overlooked.

LACTIC ACID BACTERIA

The lactic acid bacteria are a group of Gram positive bacteria, non-respiring, non-spore forming, cocci or rods, which produce lactic acid as the major end product of the fermentation of carbohydrates. They are the most important bacteria in desirable food fermentations, being responsible for the fermentation of sour dough bread, sorghum beer, all fermented milks, cassava (to produce gari and fufu) and most "pickled" (fermented) vegetables.

Historically, bacteria from the genera Lactobacillus, Leuconostoc, Pediococcus and Streptococcus are the main species involved. Several more have been identified, but play a minor role in lactic fermentations. Lactic acid bacteria were recently reviewed by Axelsson.

Lactic acid bacteria carry out their reactions - the conversion of carbohydrate to lactic acid plus carbon dioxide and other organic acids - without the need for oxygen. They are described as microaeropHilic as they do not utilise oxygen. Because of this, the changes that they effect do not cause drastic changes in the composition of the food. Some of the family are homofermentative, that is they only produce lactic acid, while others are heterofermentative and produce lactic acid plus other volatile compounds and small amounts of alcohol. Lactobacillus acidopHilus, L. bulgaricus, L. plantarum, L. caret, L. pentoaceticus, L brevis and L. thermopHilus are examples of lactic acid-producing bacteria involved in food fermentations.

All species of lactic acid bacteria have their own particular reactions and niches, but overall, L. plantarum - a homofermenter -produces high acidity in all vegetable fermentations and plays the major role. All lactic acid producers are non-motile gram positive rods that need complex carbohydrate substrates as a source of energy.

The lactic acid they produce is effective in inhibiting the growth of other bacteria that may decompose or spoil the food. Because the whole groups are referred to as 'lactic acid bacteria' it might appear that the reactions they carry out are very simple, with the production of one substrate. This is far from the truth. The lactic acid bacteria are a diverse group of organisms with a diverse metabolic capacity. This diversity makes them very adaptable to a range of conditions and is largely responsible for their success in acid food fermentations.

Despite their complexity, the whole basis of lactic acid fermentation centres on the ability of lactic acid bacteria to produce acid, which then inhibits the growth of other non-desirable organisms. All lactic acid producers are micro-aeropHilic, which is they require small amounts of oxygen to function. Species of the genera Streptococcus and Leuconostoc produce the least acid.

Next are the heterofermentative species of Lactobacillus which produce intermediate amounts of acid, followed by the Pediococcus and lastly the homofermenters of the Lactobacillus species, which produce the most acid. Homofermenters, convert sugars primarily to lactic acid, while heterofermenters produce about 50% lactic acid plus 25% acetic acid and ethyl alcohol and 25% carbon dioxide. These other compounds are important as they impart particular tastes and aromas to the final product. The heterofermentative lactobacilli produce mannitol and some species also produce dextran.

Leuconostoc mesenteroides is a bacterium associated with the sauerkraut and pickle fermentations. This organism initiates the desirable lactic acid fermentation in these products. It differs from other lactic acid species in that it can tolerate fairly high concentrations of salt and sugar (up to 50% sugar). L. mesenteroides initiates growth in vegetables more rapidly over a range of temperatures and salt concentrations than any other lactic acid bacteria.

It produces carbon dioxide and acids which rapidly lower the pH and inhibit the development of undesirable microorganisms. The carbon dioxide produced replaces the oxygen, making the environment anaerobic and suitable for the growth of subsequent species of lactobacillus. Removal of oxygen also helps to preserve the colour of vegetables and stabilises any ascorbic acid that is present.

Organisms from the gram positive Propionibacteriaceae family are responsible for the flavour and texture of some fermented foods, especially Swiss cheese, where they are responsible for the formation of 'eyes' or holes in the cheese. These bacteria break down lactic acid into acetic and propionic acids and carbon dioxide.

Several other bacteria, for instance Leuconostoc citrovorum L. Dextranicum, Streptococcus lactis, S. Cremis, & liquefaciens and Brevibacterium species are important in the fermentation of dairy products. They are not discussed in detail in this manuscript.

LACTIC ACID FERMENTATION

The lactic acid bacteria belong to two main groups - the homofermenters and the heterofermenters. The pathways of lactic acid production differ for the two. Homofermenters produce mainly lactic acid, via the glycolytic. Heterofer menters produce lactic acid plus appreciable amounts of ethanol, acetate and carbon dioxide, via the 6-pHospHogl ucanate/pHospHoketolase pathway.

The glycolytic pathway is used by all lactic acid bacteria except leuconostocs, group III lactobacilli, oenococci and weissellas. Normal conditions required for this pathway are excess sugar and limited oxygen. Axelsson gives an in-depth account of the biochemical pathways for both homo- and hetero-fermenters.

Homolactic Fermentation

The fermentation of 1 mole of glucose yields two moles of lactic acid;

$C_6H_{12}O_6 \rightarrow$	$2CH_3CHOHCOOH$
Glucose	lactic acid

$C_6H_{12}O_6 \rightarrow$	$CH_3CHOHCOOH+$	C_2H_5OH+	CO_2
Glucose	lactic acid+	ethanol+	carbon dioxide

ACETIC ACID BACTERIA

A second group of bacteria of importance in food fermentations are the acetic acid producers from the Acetobacter species. Acetobacter are important in the production of vinegar (acetic acid) from fruit juices and alcohols. The same reaction also occurs in wines, oxygen permitting, where the acetobacter can cause undesirable changes the oxidation of alcohol to acetic acid. This produces a vinegary off-taste in the wine.

The most desirable action of acetic acid bacteria is in the production of vinegar. The vinegar process is essentially a two stage process, where yeasts convert sugars into alcohol, followed by acetobacter, which oxidise alcohol to acetic acid.

BACTERIA OF ALKALINE FERMENTATIONS

A third group of bacteria are those which bring about alkaline fermentations the Bacillus species. Of note are Bacillus subtilis, B. licheniformis and B. pumilius. Bacillus subtilis is the dominant species, causing the hydrolysis of protein to amino acids and peptides and releasing ammonia, which increases the alkalinity and makes the substrate unsuitable for the growth of spoilage organisms. Alkaline fermentations are more common with protein rich foods such as soybeans and other legumes, although there are a few examples utilising plant seeds. For example water melon seeds and sesame seeds and others where coconut and leaf proteins are the substrates.

Heterolactic Fermentation

Major Lactic Acid Bacteria in Fermented Plant Products.

Homofermenter	Facultative homofermenter	Obligate heterofermenter
Enterococcus faecium	Lactobacillus bavaricus	Lactobacillus brevis
Enterococcus faecalis	Lactobacillus casei	Lactobacillus buchneri
Lactobacillus acidopHilus	Lactobacillus coryniformis	Lactobacillus cellobiosus
Lactobacillus lactis	Lactobacillus curvatus	Lactobacillus confusus
Lactobacillus delbrueckii	Lactobacillus plantarum	Lactobacillus copropHilus
Lactobacillus leichmannii	Lactobacillus sake	Lactobacillus fermentatum
Lactobacillus salivarius		Lactobacillus sanfrancisco
Streptococcus bovis		Leuconostoc dextranicum
Streptococcus thermopHilus		Leuconostoc mesenteroides
Pediococcus acidilactici		Leuconostoc paramesenteroides
Pedicoccus damnosus		
Pediococcus pentocacus		

Oxidation of Alcohol to Acetic Acid and Water

The oxidation of one mole of ethanol yields one mole each of acetic acid and water;

C_5H_5OH+	$O_2\rightarrow$	CH_3COOH+	H_2O
Alcohol		acetic acid	water

Although the range of products of alkaline fermentation does not match those brought about by acid fermentations, they are important in that they provide protein rich, low cost condiments from leaves, seeds and beans, which contribute to the diet of millions of people in Africa and Asia. Steinkraus presents a comprehensive review of the acid, alkaline and alcoholic fermentations from around the world.

CONDITIONS REQUIRED FOR BACTERIAL FERMENTATIONS

Micro-organisms vary in their optimal pH requirements for growth. Most bacteria favour conditions with a near neutral pH (7). The varied pH requirements of different groups of micro-organisms is used to good effect in fermented foods where successions of micro-organisms take over from each other as the pH of the environment changes. Certain bacteria are acid tolerant and will survive at reduced pH levels. Notable acid-tolerant bacteria include the Lactobacillus and Streptococcus species, which play a role in the fermentation of dairy and vegetable products.

Oxygen requirements vary from species to species. The lactic acid bacteria are described as microaeropHilic as they carry out their reactions with very little oxygen. The acetic acid bacteria however, require oxygen to oxidise alcohol to acetic acid. In vinegar production, oxygen has to be made available for the production of acetic acid, whereas with wine it is essential to exclude oxygen to prevent oxidation of the alcohol and spoilage of the wine.

Temperature

Different bacteria can tolerate different temperatures, which provide enormous scope for a range of fermentations. While most bacteria have a temperature optimum of between 20 to 30°C, there are some (the thermopHiles) which prefer higher temperatures (50 to 55°C) and those with colder temperature optima (15 to 20°C). Most lactic acid bacteria work best at temperatures of 18 to 22°C. The Leuconostoc species which initiate fermentation have an optimum of 18 to 22°C. Temperatures above 22°C, favour the lactobacillus species.

Salt Concentration

Lactic acid bacteria tolerate high salt concentrations. The salt tolerance gives them an advantage over other less tolerant species and allows the lactic acid fermenters to begin metabolism, which produces acid, which further inhibits the growth of non-desirable organisms. Leuconostoc is noted for its high salt tolerance and for this reason, initiates the majority of lactic acid fermentations.

Water Activity

In general, bacteria require a fairly high water activity (0.9 or higher) to survive. There are a few species which can tolerate water activities lower than this, but usually the yeasts and fungi will predominate on foods with a lower water activity.

Hydrogen ion Concentration (pH)

The optimum pH for most bacteria is near the neutral point (pH 7.0). Certain bacteria are acid tolerant and will survive at reduced pH levels. Notable acid-tolerant bacteria include the Lactobacillus and Streptococcus species, which play a role in the fermentation of dairy and vegetable products.

OXYGEN AVAILABILITY

Some of the fermentative bacteria are anaerobes, while others require oxygen for their metabolic activities. Some, lactobacilli in particular, are microaeropHilic. That is they grow in the presence of reduced amounts of atmospHeric oxygen. In aerobic fermentations, the amount of oxygen present is one of the limiting factors. It determines the type and amount of biological product obtained the amount of substrate consumed and the energy released from the reaction. Acetobacter require oxygen for the oxidation of alcohol to acetic acid.

Nutrients

All bacteria require a source of nutrients for metabolism. The fermentative bacteria require carbohydrates either simple sugars such as glucose and fructose or complex carbohydrate, such as starch or cellulose. The energy requirements of micro-organisms are very high. Limiting the amount of substrate available can check their growth.

Principles of Lactic Acid Fermentation

Sauerkraut is one example of an acid fermentation of vegetables. The name sauerkraut literally translates as acid cabbage. The 'sauerkraut process' can be applied to any other suitable type of vegetable product. Because of the importance of this product in the German diet, the process has received substantial research in order to commercialise and standardise production. As a result, the process and the contributing microorganisms are known intimately.

Other less well known fermented fruits and vegetables have received less research attention, therefore little is known of the exact process. It is safe to assume however that the acid fermentation of vegetables is based on this process. Lactic acid fermentations are carried out under three basic types of condition: dry salted, brined and non-salted. Salting provides a suitable environment for lactic acid bacteria to grow which impart the acid flavour to the vegetable.

SALT FOR PICKLING.

For pickling any variety of common salt is suitable as long as it is pure. Impurities or additives can cause problems. Salt with chemicals to reduce caking should not be used as they make the brine cloudy. Salt with lime impurities can reduce the acidity of the final product and reduce the shelf life of the product. Salt with iron impurities can result in the blackening of the vegetables. Magnesium impurities impart a bitter taste. Carbonates can result in pickles with a soft texture.

DRY SALTED FERMENTED VEGETABLES

With dry salting, the vegetable is treated with dry salt. The salt extracts the juice from the vegetable and creates the brine. The vegetable is prepared, washed in potable cold water and drained. For every 100 kg of vegetables 3 kg of salt is needed. The vegetables are placed in a layer of about 2.5cm depth in the fermenting container (a barrel or keg). Salt is sprinkled over the vegetables.

Another layer of vegetables is added and more salt added. This is repeated until the container is three quarters full. A cloth is placed above the vegetables and a weight added to compress the vegetables and assist the formation of brine which takes about 24 hours. As soon as the brine is formed, fermentation starts and bubbles of carbon dioxide begin to appear. Fermentation takes between one and four weeks depending on the ambient temperature. Fermentation is complete when no more bubbles appear, after which time the pickle can be packaged in a variety of mixtures. These can be vinegar and spices or oil and spices.

THE FERMENTATION PROCESS

Shredded cabbage or other suitable vegetables are placed in a jar and salt is added. Mechanical pressure is applied to the cabbage to expel the juice, which contains fermentable sugars and other nutrients suitable for microbial activity. The first micro-organisms to start acting are the gas-producing cocci (L. Mesenteroides).

These microbes produce acids. When the acidity reaches 0.25 to 0.3% (calculated as lactic acid), these bacteria slow down and begin to die off, although their enzymes continue to function. The activity initiated by the L. mesenteroides is continued by the lactobacilli (L. plantarum and L. Cucumeris) until an acidity level of 1.5 to 2% is attained. The high salt concentration and low temperature inhibit these baicteria to some extent. Finally, L. pentoaceticus continues the fermentation, bringing the acidity to 2 to 2.5% thus completing the fermentation.

The end products of a normal kraut fermentation are lactic acid along with smaller amounts of acetic and propionic acids, a mixture of gases of which carbon dioxide is the principal gas, small amounts of alcohol and a mixture of aromatic esters. The acids, in combination with alcohol form esters, which contribute to the characteristic flavour of sauerkraut. The acidity helps to control the growth of spoilage and putrefactive organisms and contributes to the extended shelf life of the product. Changes in the sequence of desirable bacteria, or indeed the presence of undesirable bacteria, alter the taste and quality of the product.

EFFECTS OF TEMPERATURE ON SAUERKRAUT PROCESS

The optimum temperature for sauerkraut fermentation is around 21°C. A variation of just a few degrees from this temperature alters the activity of the microbial process and affects the quality of the final product. Therefore, temperature control is one of the most important factors in the sauerkraut process. A temperature of 18° to 22° C is most desirable for initiating fermentation since this is the optimum temperature range for the growth and metabolism of L. mesenteroides. Temperatures above 22°C favour the growth of Lactobacillus species.

EFFECTS OF SALT ON THE SAUERKRAUT PROCESS

Salt plays an important role in initiating the sauerkraut process and affects the quality of the final product. The addition of too much salt may inhibit the desirable bacteria, although it may contribute to the firmness of the kraut. The principle function of salt is to withdraw juice from the cabbage (or other vegetable), thus making a more favourable environment for development of the desired bacteria.

Generally, salt is added to a final concentration of 2.0 to 2.5%. At this concentration, lactobacilli are slightly inhibited, but cocci are not affected. Unfortunately, this concentration of salt has a greater inhibitory effect against the desirable organisms than against those responsible for spoilage. The spoilage organisms can tolerate salt concentrations up to between 5 and 7%, therefore it is the acidic environment created by the lactobacilli that keep the spoilage bacteria at bay, rather than the addition of salt.

In the manufacture of sauerkraut, dry salt is added at the rate if 1 to 1.5 kg per 50kg cabbage (2 to 3%). The use of salt brines is not recommended in sauerkraut making, but is common in vegetables that have a low water content. It is essential to use pure salt since salts with added alkali may neutralise the acid.

USE OF STARTER CULTURES

In order to produce sauerkraut of consistent quality, starter cultures (similar to those used in the dairy industry) have been recommended. Not only do starter cultures ensure consistency between batches, they speed up the fermentation process as there is no time lag while the relevant microflora colonise the sample. Because the starter cultures used are acidic, they also inhibit the undesirable micro-organisms.

It is possible to add starters traditionally used for milk fermentation, such as Streptococcus lactis, without adverse effect on final quality. Because these organisms only survive for a short time (long enough to initiate the acidification process) in the kraut medium, they do not disturb the natural sequence of micro-organisms.

On the other hand, if Leuconostoc mesenteroides is added in the early stages, it gives a good flavour to the final product, but alters the sequence of subsequent bacterial growth and results in a product that is incompletely fermented. If gas producing rods (for example L pentoaceticus) are added to the sauerkraut, this disturbs the balance between acetic and lactic acids - more acetic acid and less lactic acid are produced than normal - and the fermentation never reaches completion. If lactic acids, non-gas producing rods (L. Cucumeris) are used as a starter, again the kraut is not completely fermented and the resulting product is bitter and more susceptible to spoilage by yeasts.

It is possible to use the juice from previous kraut fermentation as a starter culture for subsequent fermentations. The efficacy of using old juice depends largely on the types of organisms present in the juice and its acidity. If the starter juice has an acidity of 0.3% or more, it results in a poor quality kraut. This is because the cocci which would normally initiate fermentation are suppressed by the high acidity, leaving the bacilli with sole responsibility for fermentation. If the starter juice has an acidity of 0. 25% or less, the kraut produced is normal, but there do not appear to be any beneficial effects of adding this juice. Often, the use of old juice produces sauerkraut which has a softer texture than normal.

SPOILAGE AND DEFECTS IN THE SAUERKRAUT PROCESS.

The majority of spoilage in sauerkraut is due to aerobic soil micro-organisms which break down the protein and produce undesirable flavour and texture changes. The growth of these aerobes can easily be inhibited by a normal fermentation.

Soft kraut can result from many conditions such as large amounts of air, poor salting procedure and varying temperatures. Whenever the normal sequence of bacterial growth is altered or disturbed, it usually results in a soft product. It is the lactobacilli, which seem to have a greater ability than the cocci to break down cabbage tissues, which are responsible for the softening. High temperatures and a reduced salt content favour the growth of lactobacilli, which are sensitive to higher concentrations of salt. The usual concentration of salt used in sauerkraut production slightly inhibits the lactobacilli, but has no effect on the cocci. If the salt content is too low initially, the lactobacilli grow too rapidly at the beginning and upset the normal sequence of fermentation.

Another problem encountered is the production of dark coloured sauerkraut. This is caused by spoilage organisms during the fermentation process. Several conditions favour the growth of spoilage organisms. For example, an uneven distribution of salt tends to inhibit the desirable organisms while at the same time allowing the undesirable salt tolerant organisms to flourish. An insufficient level of juice to cover the kraut during the fermentation allows undesirable aerobic bacteria and yeasts to grow on the surface of the kraut, causing off flavours and discoloration. If the fermentation temperature is too high, this also encourages the growth of undesirable microflora, which results in a darkened colour.

Pink kraut is a spoilage problem. It is caused by a group of yeasts which produce an intense red pigment in the juice and on the surface of the cabbage. It is caused by an uneven distribution of or an excessive concentration of salt, both of which allow the yeast to multiply. If conditions are optimal for normal fermentation, these spoilage yeasts are suppressed.

BRINE SALTED FERMENTED VEGETABLES

Brine is used for vegetables which inherently contain less moisture. A brine solution is prepared by dissolving salt in water (a 15 to 20% salt solution). Fermentation takes place well in a brine of about 20 salometer. As a general guide, a fresh egg floats in a 10% brine solution. Properly brined vegetables will keep well in vinegar for a long time.

The duration of brining is important for the overall keeping qualities. The vegetable is immersed in the brine and allowed to ferment. The strong brine solution draws sugar and water out of the vegetable, which decreases the salt concentration. It is crucial that the salt concentration does not fall below 12%, otherwise conditions do not allow for fermentation. To achieve this, extra salt is added periodically to the brine mixture.

Once the vegetables have been brined and the container sealed, there is a rapid development of micro-organisms in the brine. The natural controls which affect the microbial populations of the fermenting vegetables include the concentration of salt and temperature of the brine, the availability of fermentable materials and the numbers and types of micro-organisms present at the start of fermentation.

The rapidity of the fermentation is correlated with the concentration of salt in the brine and its temperature.

Most vegetables can be fermented at 12.5° to 20° salometer salt. If so, the microbial sequence of lactic acid bacteria generally follows the classical sauerkraut fermentation described by Pederson. At higher salt levels of up to about 40° salometer, the sequence is skewed towards thedevelopment of a homofermentation, dominated by Lactobacillus plantarum. At the highest concentrations of salt (about 60° salometer) the lactic fermentation ceases to function and if any acid is detected during brine storage it is acetic acid, presumably produced by acid-forming yeasts which are still active at this concentration of salt.

BRINE SALTED FERMENTATION OF VEGETABLES (PICKLES)

Pickled cucumbers are another fermented product that has been studied in detail and the process is known. The fermentation process is very similar to the sauerkraut process, only brine is used instead of dry salt. The washed cucumbers are placed in large tanks and salt brine (15 to 20%) is added. The cucumbers are submerged in the brine, ensuring that none float on the surface - this is essential to prevent spoilage.

The strong brine draws the sugar and water out of the cucumbers, which simultaneously reduces the salinity of the solution. In order to maintain a salt solution so that fermentation can take place, more salt has to be added to the brine solution. If the concentration of salt falls below 12%, it will result in spoilage of the pickles through putrefaction and softening.

A few days after the cucumbers have been placed in the brine, the fermentation process begins. The process generates heat which causes the brine to boil rapidly. Acids are also produced as a result of the fermentation. During fermentation, visible changes take place which are important in judging the progress of the process. The colour of the cucumber surface changes from bright green to a dark olive green as acids interact with the chloropHyll. The interior of the cucumber changes from white to a waxy translucent shade as air is forced out of the cells. The specific gravity of the cucumbers also increases as a result of the gradual absorption of salt and they begin to sink in the brine rather than floating on the surface.

MICROBES INVOLVED IN THE FERMENTATION PROCESS

As with the sauerkraut process, the gram positive coccus Leuconostoc mesenteroides predominates in the first stages of pickle fermentation. This species is more resistant to temperature changes and tolerates higher salt concentration than the subsequent species.

As fermentation proceeds and the acidity increases, lactobacilli start to take over from the cocci. The active stage of fermentation continues for between 10 to 30 days, depending upon the temperature of the fermentation. The optimum temperature for L. Cucumeris is 29 to 32°C. During the fermentative period, the acidity increases to about 2% and the strong acid producing types of bacteria reach their maximum growth. If sugar or acetic acid is added to the fermenting mixture during this time it increases the production of acid.

PROBLEMS IN PICKLES

The production of excessive amounts of acid during the fermentation, results in shrivelling of the pickles, possibly due to over-activity of the L. mesenteroides species. If the brine is stirred, it may introduce air, which makes conditions more favourable for the growth of spoilage bacteria. In general, if the pickles are well covered with brine, the salt concentration is maintained and the temperature is at an optimum, it should be quite simple to produce good quality pickles.

NON SALTED, LACTIC ACID FERMENTED VEGETABLES

Some vegetables are fermented by lactic acid bacteria, without the prior addition of salt or brine. Examples of non-salted products include gundruk (consumed in Nepal), sinki and other wilted fermented leaves. The detoxification of cassava through fermentation includes an acid fermentation, during which time the cyanogenic glycosides are hydrolysed to liberate the toxic cyanide gas.

The fermentation process relies on the rapid colonisation of the food by lactic acid producing bacteria, which lower the pH and make the environment unsuitable for the growth of spoilage organisms. Oxygen is also excluded as the Lactobacilli favour an anaerobic atmospHere. Restriction of oxygen ensures that yeasts do not grow.

For the production of sinki, fresh radish roots are harvested, washed and wilted by sun-drying for one to two days. They are then shredded, re-washed and packed tightly into an earthenware or glass jar, which is sealed and left to ferment. The optimum fermentation time is twelve days at 30°C. Sinki fermentation is initiated by L. fermentum and L. brevis, followed by L. plantarum.

During fermentation the pH drops from 6. 7 to 3. 3. After fermentation, the radish substrate is sun-dried to a moisture level of about 21%. For consumption, sinki is rinsed in water for two minutes, squeezed to remove the excess water and fried with salt, tomato, onion and green chilli. The fried mixture is then boiled in rice water and served hot as soup along with the main meal.

PIT FERMENTATIONS

South Pacific pit fermentations are an ancient method of preserving starchy vegetables without the addition of salt. The raw materials undergo an acid fermentation within the pit, to produce a paste with good keeping qualities. Pit fermentations are also used in other parts of the world - for example in Ethiopia, where the false banana (Ensete ventricosum) is fermented in a pit to produce a pulp known as kocho. Foods preserved in pits can last for years without deterioration, therefore pits provide a good, reliable cheap means of storage.

Root crops and bananas are peeled before being placed in the pit, while breadfruit are scraped and pierced. Food is left to ferment for three to six Weeks, after which time it becomes soft, has a strong odour and a paste-like consistency. During fermentation, carbon dioxide builds up in the pit, creating an anaerobic atmospHere. As a result of bacterial activity, the temperature rises much higher than the ambient temperature.

The pH of the fruit within the pit decreases from 6. 7 to 3. 7 within about four weeks. Inoculation of the fruit in the pit with lactic acid bacteria greatly speeds up the process. The fermented paste can be left in the pit and removed as required. Usually, it is removed and replaced with a second batch of fresh food to ferment. The fermented food is washed and fibrous material removed. It is then dried in the sun for several hours to remove the volatile odours, and pounded into a paste. Grated coconut or coconut cream and sugar may be added and the mixture is wrapped in banana leaves and either baked or boiled.

PRINCIPLES OF ACETIC ACID FERMENTATION

The main desirable fermentation carried out by acetic acid bacteria is the production of vinegar. Vinegar, literally translated as sour wine, is one of the oldest products of fermentation used by man. It can be made from almost any fermentable carbohydrate source, for example fruits, vegetables, syrups and wine. Whatever the raw material used, the fermentation process follows a definite sequence.

The basic requirement for vinegar production is a raw material that will undergo an alcoholic fermentation. Apples, pears, grapes, honey, syrups, cereals, hydrolysed starches, beer and wine are all ideal substrates for the production of vinegar. The best raw materials are cider and wine, which are widely used in Europe and the United States. To produce a high quality product it is essential that the raw material is mature, clean and in good condition.

MICROBES INVOLVED IN THE VINEGAR PROCESS.

The production of vinegar depends on a mixed fermentation, which involves both yeasts and bacteria. The fermentation is usually initiated by yeasts which break down glucose into ethyl alcohol with the liberation of carbon dioxide gas. Following on from the yeasts, acetobacter oxidise the alcohol to acetic acid and water.

Yeast reaction			
$C_6H_{12}O_6$	$\longrightarrow$	$2C_2H_5OH +$	$2CO_2$
Glucose	yeast	ethyl alcohol +	carbon dioxide
Bacterial reaction			
$C_2H_5OH + O_2$	$\longrightarrow$	$CH_3COOH +$	H_2O
Alcohol		acetic acid	water

The yeasts and bacteria exist together in a form known as commensalism. The acetobacter are dependent upon the yeasts to produce an easily oxidisable substance (ethyl alcohol). It is not possible to produce vinegar by the action of one type of micro-organism alone.

For a good fermentation, it is essential to have an alcohol concentration of 10 to 13%. If the alcohol content is much higher, the alcohol is incompletely oxidised to acetic acid. If it is lower than 13%, there is a loss of vinegar because the esters and acetic acid are oxidised. In addition to acetic acid, other organic acids are formed during the fermentation which become esterified and contribute to the characteristic odour, flavour and colour of the vinegar.

Acetaldehyde is an intermediate product in the transformation of the reducing sugar in fruit juice to acetic acid or vinegar. Oxygen is required for the conversion of acetaldehyde to acetic acid. In general, the yield of acetic acid from glucose is approximately 60%. That is three parts of glucose yield two parts acetic acid.

MICRO-ORGANISMS INVOLVED IN THE FERMENTATION OF VINEGAR.

The organisms involved in vinegar production usually grow at the top of the substrate, forming a jelly like mass. This mass is known as 'mother of vinegar'. The mother is composed of both acetobacter and yeasts, which work together. The principal bacteria are Acetobacter acetic A. Xylinum and A. Ascendens. The main yeasts are Saccharomyces ellipsoideus and S cerevisiae. It is important to maintain an acidic environment to suppress the growth of undesirable organisms and to encourage the presence of desirable acetic acid producing bacteria. It is common practice to add 10 to 25% by volume of strong vinegar to the alcoholic substrate in order to attain a desirable fermentation.

The alcoholic fermentation of sugars should be completed before the solution is acidified because any remaining sugar will not be converted to alcohol after the acetic acid is added. Incomplete fermentation of the juice results in a "weak" product. The acetic acid strength of good vinegar should be approximately 6%.

FERMENTATION METHODS

Small Scale Production

Vinegar can be made at home at the small scale by introducing oxygen into barrels of wine or cider and allowing fermentation to occur spontaneously. This process is not very rigorously controlled and often results in a poor quality product.

The Orleans Process

The Orleans process is one of the oldest and well known methods for the production of vinegar. It is a slow, continuous process, which originated in France. High grade vinegar is used as a starter culture, to which wine is added at weekly intervals. The vinegar is fermented in large (200 litre) capacity barrels. Approximately 65 to 70 litres of high grade vinegar is added to the barrel along with 15 litres of wine. After one week, a further 10 to 15 litres of wine are added and this is repeated at weekly intervals. After about four weeks, vinegar can be withdrawn from the barrel (10 to 15 litres per week) as more wine is added to replace the vinegar.

One of the problems encountered with this method is that of how to add more liquid to the barrel without disturbing the floating bacterial mat. This can be overcome by using a glass tube which reaches to the bottom of the barrel. Additional liquid is poured in through the tube and therefore does not disturb the bacteria. Wood shavings are sometimes added to the fermenting barrel to help support the bacterial mat.

Quick Vinegar Method

Because the Orleans process is slow, other methods have been adapted to try and speed up the process. The German method is one such method. It uses a generator, which is an upright tank filled with beechwood shavings and fitted with devices which allow the alcoholic solution to trickle down through the shavings in which the acetic acid bacteria are living. The tank is not allowed to fill as that would exclude oxygen which is necessary for the fermentation. Near the bottom of the generator are holes which allow air to be drawn in. the air rises through the generator and is used by the acetic acid bacteria to oxidise the alcohol. This oxidisation also releases considerable amounts of heat which must be controlled to avoid causing damage to the bacteria.

Problems in Vinegar Production

Many of the problems of vinegar production are concerned with the presence of nematodes, mites, flies and other insects. These pests can be controlled by adherence to good hygiene and pasteurisation of the vinegar. Problems associated with the fermentative process include the presence of a whitish film on the surface of the vinegar.

This is sometimes called Mycoderma vini and is composed of yeast-like organisms, which grow aerobically and oxidise the carbon containing compounds to carbon dioxide and water. They also alter the flavour and alcohol content of the vinegar. This problem can however be controlled by adding one part vinegar to three parts of the alcoholic solution or by storing the alcoholic liquid in filled closed containers.

TYPICAL PRODUCTS OF BACTERIAL FERMENTATIONS

Many products are produced by bacterial fermentations. These include the fruit and vegetable pickles produced by lactic acid fermentation and the products of alkaline bacterial fermentations. Lactic acid bacteria pickling is still carried out at the domestic scale. However industrial scale processes have been developed for the production of most types of pickles. Pickles can be made by storing prepared vegetables in a weak brine solution, by dry salting or allowing the vegetables to ferment without salt.

DRY SALTED PICKLES

Dry salting is used for pickling many vegetables and fruits including limes, lemons and cucumbers. For dry salt pickling any variety of common salt is suitable as long as it is pure. Impurities or additives can cause problems:

- Chemicals to reduce caking should not be used as they make the brine cloudy.
- Lime impurities can reduce the acidity of the final product and reduce the shelf life of the product.
- Iron impurities can result in the blackening of the vegetables.
- Magnesium impurities impart a bitter taste.
- Carbonates can result in pickles with a soft texture.

DRY SALTED LIME PICKLE

Product Description

With dry salting, the limes are treated with dry salt. The salt extracts and juice from the vegetable and create the brine. The final product is a sour lime pickle. Spices are added depending on local preference. In India and Pakistan, the pickle is usually very spicy and hot due to the addition of chilli. It is usually eaten as a condiment.

Preparation of Raw Materials

The limes need to be selected and prepared. Only fully ripe limes without bruising or damage should be used. All limes need to be washed in potable cold water, drained, and then cut into quarters. Spices should be of good quality and free of mould.

Processing

Limes are placed in a layer, approximately 2.5 cm deep, into the fermenting container (a barrel or keg). One kilogram of salt is added for every four kilograms of limes. The salt is sprinkled over the vegetables. Another layer of vegetables is added and more salt added. This is repeated until the container is three quarters full. A cloth is placed above the vegetables and a weight added to compress the vegetables and assist in the formation of a brine. The formation of a brine takes about 24 hours. The container is then placed in the sun for a week.

As soon as the brine is formed, fermentation starts. As fermentation starts bubbles of carbon dioxide appear. Fermentation takes between one and four weeks depending on the ambient temperature. Fermentation is complete when no more bubbles appear.

Flow Diagram

Selection ↓	Only ripe limes should be selected
Wash ↓	In clean water
Cut ↓	Cut into four pieces or slice the skin
Mix with salt ↓	1kg salt for 4kg of limes
Ferment ↓	Leave the container in the sun for a week to ferment
Package ↓	

Packaging and Storage

The vegetables can be removed from the brine and packaged in a variety of mixtures which may consist of vinegar and spices or oil and spices. Lime pickle can be packed in small polythene bags and sealed or in clean jars and capped. Lime pickles keep well if stored in a cool place. Due to the high acid level of the final product, the risk of food poisoning is low. -

PICKLED CUCUMBERS

Product Description

Cucumbers undergo a typical lactic acid fermentation and change from a pale product to a darker green and more transparent product. Khalpi is a cucumber pickle popular during the summer months in Nepal.

Fully ripe cucumbers without bruising or damage are washed in potable cold water and drained. The cucumbers can be pickled whole or sliced. With khalpi the cucumbers are washed, sliced and cut into 5-8 cm pieces.

Flow Diagram

Selection ↓	Only ripe cucumbers should be selected
Wash ↓	In clean water
Mix with salt ↓	1kg salt for 15-20kg of cucumbers
Ferment ↓	For between one and four weeks
Package ↓	

Processing

1 kg of salt is added to every 20 kg of small cucumbers and 15 kg of large cucumbers. The brine should be formed within 24 hours by osmosis. If the brine formed by osmosis does not cover the cucumbers 40o Salometer brine is added to the desired level. A day or two after the tank is filled and closed the brine should be stirred in order to help equalise the concentration of salt throughout the mass. As soon as the brine is formed, fermentation starts and bubbles of carbon dioxide appear.

Fermentation takes between one and four weeks depending on the ambient temperature. Fermentation is complete when no more bubbles appear. During fermentation the brine becomes cloudy for the first few days due to the growth of bacteria. Later if the brine is not covered, a filmy yeast growth will often occur on the surface.

Packaging and Storage

Cucumber pickle is usually stored in clean capped jars. They keep well if stored in a cool place. Due to the high acid level of the final product, the risk of food poisoning is low. With khalpi in Nepal, oil is added.

PAK-GARD-DONG (PICKLED LEAFY VEGETABLE)

Pak-Gard-Dong is a fermented mustard leaf (Brassica juncea) product made in Thailand. The mustard leaves are washed, wilted in the sun, mixed with salt, packed into containers for 12 hours. The water is then drained and a 3% sugar solution added. They are again allowed to ferment for three to five days at room temperature. Micro-organisms associated with the fermentation include Lactobacillus brevis, Pediococcus cerevisiae and Lactobacillus plantarum.

A similar product (Hum choy) is made in the South of China. This is produced by fermenting a local leafy vegetable. The leaves are washed and drained. They are then covered in salt and hung on racks to dry in the sun. The wilted leaves are placed in earthenware pots and covered with rice water, obtained after washing rice grains. The pots are sealed and the leaves allowed to ferment for four days. The product can be stored for up to two months if the seal is not broken.

Tempoyak (Pickled Durian)

Tempoyak is the fermented pulp of a durian fruit (Durio zibethinus) from Malaysia. It has the distinctive durian smell and a creamy yellow colour. It is made by mixing durian pulp with salt and placing in a sealed container. Fermentation takes about seven days.

Pickled Beetroots

In Russia beetroot is pickled by cleaning, slicing and placing in a container with salt. Due to the high sucrose level, dextrans are produced giving the product a slimy texture.

Lamoun Makbous (Pickled Lemons)

Pickled lemons are popular in Asia. In West Asia and north Africa they are known as lamoun makbous and msir. Lemons are washed in clean water, sliced and covered in salt. After at least 24 hours, they are drained and mixed with oil and spices.

BRINED FRUIT AND VEGETABLE PICKLES

For brine pickling any variety of common salt is suitable as long as it is pure. Impurities or additives can cause problems:

- Chemicals to reduce caking should not be used as they make the brine cloudy.
- Lime impurities can reduce the acidity of the final product and reduce the shelf life of the product.
- Iron impurities can result in the blackening of the vegetables.
- Magnesium impurities impart a bitter taste.
- Carbonates can result in pickles with a soft texture.

GREEN MANGO PICKLE

Product Description

Green mango pickle is a hot, spicy pickle with a sour taste. It is eaten as a condiment. Preservation is caused by a combination of salt, increased acidity and to a small extent the spices. It is known as burong mangga and dalok in the Philippines.

Preparation of the Raw Material

The fresh, fully mature, firm but unripe mangoes must be carefully selected to ensure a good quality product. The best pickles are obtained from fruit at early maturity when the fruit has reached almost maximum size. Riper fruit results in pickles with a fruity odour and lacking the characteristic and predominant green mango flavour.

The green mangoes need to be inspected arid any damaged fruit rejected. The fruit is washed in clean water and drained. After draining, the fruit is cut. Sharp knives with preferably stainless blades should be used. Iron or copper equipment should be avoided. A single stroke should be used during the cutting process to ensure minimum damage and avoiding mushiness in the final product.

Processing

The sliced mangoes are soaked in brine solution. Sodium metabisulpHite (1000 ppm) and 1% calcium chloride are added. The containers are stored until the mangoes are pickled. The brine is then drained off and spices are mixed with the mango slices.

Flow Diagram

Fruit ↓	The mangoes need to be unripe
Sort ↓	Remove damaged fruit
Wash ↓	With clean water.
Drain ↓	
Cut ↓	Stainless steel knives should be used
Soaked in brine ↓	20% salt solution
Drain ↓	Spices to taste
Add spices ↓	
Pack ↓	Pack in containers and add oil

Packaging and Storage

The mixture is then packed and oil added onto the surface of the mixture. The mangoes should be firmly pressed down in the container. Good quality vegetable oil such as sunflower oil should be used and finely ground chilli powder can be added to the oil for flavour and colour. Mango pickle can be packed in small polythene bags and sealed or in clean jars and capped. Mango pickle keeps well if stored in a cool place. If it is processed well, it can be kept for several months. Due to the high acid level of the final product, the risk of food poisoning is low.

LIME PICKLE (BRINED)

Product Description

Lime pickle is made from salted pieces of lime packed in a salty, spicy liquor, like a semi-solid gravy. It is brownish red and the lime peels are yellow or pale green with a sour and salty taste. It is eaten as a condiment with curries or other main meals. If processed well, the product can be kept for several months.

Preparation of Raw Materials

The limes need to be selected and prepared. Only fully ripe limes without bruising or damage should be used. All the limes need to be washed in potable cold water and drained. The limes are dipped in hot water (60°-65°C) for about five minutes. They are then cut into pieces in order to expose the interior and allow salt to be absorbed more quickly. All spices should be of good quality and free of mould.

Processing

The prepared limes are covered with a brine solution. This causes water to be drawn out of the pieces by osmosis. It is important to ensure that the surface is covered with juice, and leave for 24 hours. If necessary, the fruits should be pressed down to hold them below the liquid. Once the limes have been placed in the brine, there is a rapid development of microorganisms and fermentation begins. After fermentation the limes are dried in the sun until the skin becomes brown.

Flow Diagram

Sort ↓	Select ripe (but not overripe) healthy lime fruits
Wash ↓	
Heat ↓	Dip in hot water (60-65^{0}C) for about five minutes
Cut ↓	Cut into four pieces or alternatively cut into smaller, uniform-sized pieces
Brine ↓	Ensure that the surface is covered with juice
Dry ↓	Dry in the sun for 2-3 days
Mix spices ↓	To local preference
Pack ↓	
Store ↓	In a cool place, away from sunlight

Packaging and Storage

The limes are mixed with spices and oils according to local taste and tradition. Lime pickle can be packed in small polythene bags and sealed or in clean jars and capped. Lime pickle keeps well if stored in a cool place. Due to the high acid level of the final product, the risk of food poisoning is low.

KIMCHI (PICKLED CABBAGE)

Product Description

Kimchi is a general name for a range of closely related fermented products. It is similar to Sauerkraut in Europe and the United States. There are numerous variations of kimchi depending on the production technique. The main pickled cabbage kimchis are tongbaechu-kimchi tongkimchi and bossam-kimchi. This section refers to kimchi produced from cabbage, the following section deals with pickled radish products.

Preparation of Raw Materials

Appropriate cultivars of Chinese cabbage, with light-green coloured soft leaves and compact structures with no defects, are required for production of kimchi. After removing outer leaves and roots from the cabbage, it is cut into small pieces.

Processing

The prepared cabbage is placed in a salt solution (8-15%) for two to seven hours in order increase the salt content of the cabbage to between 2.0-4.0% (w/w). It is then rinsed several times with fresh water and drained to remove extra water by centrifugation or by allowing to stand. Kimchi fermentation is carried out by various micro-organisms present in the raw materials and ingredients used in the preparation of kimchi. Among the two hundred bacteria isolated form kimchi, the important micro-organisms in kimchi fermentation are known to be Lactobacillus plantarum, L. Brevis, Streptococcus faecalis, Leuconostoc mesenteroides and Pediococcus pentosaceus. After fermentation, the product can be left to mature for several weeks if refrigeration is available. If stored under warm conditions, the kimchi deteriorates rapidly.

GREEN OLIVES

Olives are a brine fermented product which undergo an essential pre-treatment with lye to remove substances which are toxic to bacteria and would, if left in the olives, prevent fermentation.

Green olives are placed in a 2% sodium hydroxide (lye) solution at 210 to 240°C until the lye penetrates the flesh. Cold water is added to the solution, which, dilutes the mixture until the lye is completely removed. The lye treatment is necessary to remove a bitter glucoside compound (oleuropein) from the outer tissues of the olive. Oleuropein is highly toxic to bacteria and therefore needs to be removed in order for a fermentation to take place.

Flow Diagram

Korean cabbage ↓	
Cut ↓	Cut into four parts
Add brine ↓	5% for 12 hours or 15% for 7 hours
Rinse ↓	Drain water
Mix ↓	With salt, spices and flavourings
Ferment	

After the glucoside has been removed, the olives are placed in barrels with a 1 to 10% brine solution and allowed to undergo a spontaneous fermentation. The optimum fermentation temperature is 240°C. The fermentation period usually takes between two and three months. Once fermentation is complete, the olives are packed in airtight jars and sterilised which produces a good quality product with a long storage life.

BLACK OLIVES

When ripe olives are used, they are placed in a 5 to 7% brine solution to soften the outer tissue before lye is added. This pre-treatment allows the lye to penetrate the fruit more easily. In ripe fruit, the glucoside is situated more deeply within the tissue. After soaking in brine, the lye solution (0.7 to 2.0%) is added and, after soaking, the olives are washed clean. They are then packed in barrels with a 2 to 5% brine solution and allowed to undergo fermentation for two to six weeks, depending on the external temperature.

Unlike immature olives, the ripe one are exposed to air during fermentation. This is to oxidise the polypHenols in the tissue to a black colour, which is dependent upon oxygen and the small amounts of sodium hydroxide which are left on the olives. The finished product is packed in a 3 to 5% brine solution and then sterilised. Bacillus subtilis may produce pectolytic enzymes which result in a soft product. If the wash water used for removing the lye is below 600°C, many species of undesirable bacteria can survive which result in a low quality product.

JACK-FRUIT PICKLE

Young green jack fruit is pickled in India and Sri Lanka. Young green jack fruit are peeled and cut into 1. 2 to 1. 8 cm thick slices. The slices are placed in a container and covered in an 8% common salt solution. They are weighed down to keep them submerged in the brine. The brine solution is increased by 2% each day until it reaches 15%. The slices are then left for 8-10 days in the brine. Vinegar and spices are added prior to packaging.

PICKLED RADISH

A number of pickled radish products are produced in Korea. These include: kaktugi, tongchimi, chonggak-kimchi, seokbakji, yolmu-kimchi dan moogi kach doo ki gactuki and mootsanji.

PICKLED CUCUMBER

A variety of brine pickled cucumber products are made around the world. Oi sobagi and oiji are made in Korea. In Egypt cucumbers are pickled by soaking in brine to produce torshi khiar.

PICKLED LEAFY VEGETABLES

There are many other brine pickled leafy vegetables around the world. For instance:

- Pak-sian-dong: This is a popular pickled leafy vegetable Pak Sian (Gynadropsis pentapHylla) in Thailand. The fresh vegetable is cleaned and wilted in the sun for one to two hours. It is then placed in brine and fermented for two to three days at room temperature
- Sayur asin fermented wilted mustard cabbage (Brassica juncea) from Indonesia. It is also known as kiam chai in Thailand and kiam chaye in Malaysia.

Other Pickled Vegetables and Fruits

- Naw-mai-dong: pickled bamboo shoots (Bambusa glaucescens) from Thailand
- Hom-dong: pickled red onions from Thailand
- Jeruk: pickled vegetables including ginger and papaya from Malaysia
- Pickled carrots and turnips are produced in Asia and Africa. They are known as hua-chai po in Thailand and tai tan tsoi in China.
- Nukamiso-zuke: vegetables fermented in rice bran, salt and water in Japan
- Bananas are pickled in the West Indies
- Fermented sweet peppers (torshi felfel) are produced in west Asia and Africa.

- Cauliflower stalks are fermented to produce achar tandal in India.
- Aubergines (torshi betingen) are pickled in west Asia.

NON SALTED LACTIC ACID BACTERIA PRODUCTS

GUNDRUK (PICKLED LEAFY VEGETABLE)

Product Description

Gundruk is obtained from the fermentation of leafy vegetables in Nepal. It is served as a side dish with the main meal and is also used as an appetiser. Gundruk is an important source of minerals particularly during the off-season when the diet consists of mostly starchy tubers and maize which tend to be low in minerals.

Preparation of Raw Materials

In the months of October and November, during the harvest of the first broad mustard, radish and cauliflower leaves, large quantities of leaves accumulate - much more than can be consumed fresh. These leaves are allowed to wilt for one or two days and then shredded with a knife or sickle.

Processing

Shredded leaves are tightly packed in an earthenware pot and warm water (at about 30°C) is added to cover all the leaves. The pot is then kept in a warm place. After five to seven days, a mild acidic taste indicates the end of fermentation and the gundruk is removed and sun-dried. This process is similar to sauerkraut production except that no salt is added to the shredded leaves prior to gundruk fermentation. The ambient temperature at the time of fermentation is about 18°C.

Flow Diagram

Leafy vegetables ↓	
Wilt ↓	One to two days
Shred ↓	
Placed in earthen pot ↓	The leaves need to tightly packed
Cover the leaves ↓	Cover the leaves with warm water and straw
Ferment ↓	The pot is kept warm in the sun and by a fire by night
Add warm water ↓	To keep the pot warm
Dried	Product dried on mats in the sun

Pediococcus and Lactobacillus species are the predominant micro-organisms during gundruk fermentation. The fermentation is initiated by L. cellobiosus and L. plantarum, and other homolactics make a vigorous growth from the third day onwards. Pediococcus pentosaceus increases in number on the fifth day and thereafter declines. During fermentation, the pH drops slowly to a final value of 4. 0 and the amount of acid (as lactic) increases to about 1 % on the sixth day.

It has been found that a disadvantage with the traditional process of gundruk fermentation is the loss of 90% of the carotenoids, probably during sun-drying. Improved methods of drying might reduce the vitamin loss.

KOCHO (PICKLED FALSE BANANA)

Product Description

False banana (Ensete ventricosum) is fermented in a pit to produce a pulp known as kocho. Foods preserved in pits can last for years without deterioration. Pits therefore provide a good, reliable cheap means of storage.

Processing

The type of soil and its drainage are important in the selection of a pit site. Pits are often lined with stones to prevent the soil from the side walls falling into the bottom. A family pit may be 0. 6 to 1. 5 metres deep and 1. 2 to 2 metres wide with a capacity of about fifty breadfruits. A community pit is usually much larger, with the capacity to hold up to 1000 breadfruit. A family pit requires at least 1000 green banana leaves and four sacks of dried banana leaves for lining the walls and top. It is essential that proper attention is given to hygiene of the pit and the fruit to be stored in it.

The central stems are removed from fresh banana leaves and they are wilted in the sun until they become soft and pliable. The pit is lined with dry leaves, then green leave are folded and arranged, overlapping each other, around the sides of the pit and extending over the top. At lest two or three layers of banana leaves are used to seal the pit and prevent contamination by the soil. Washed, peeled food is placed in the pit, green banana leaves are folded over the top of the food and heavy stones are placed on top to weigh down the leaves.

Root crops and bananas are peeled before placing in the pit, breadfruit are scraped and pierced. Food is left to ferment for three to six weeks, after which time it becomes soft, has a strong odour and a paste-like consistency.

The fermented paste can be left in the pit and removed as required. Usually, it is removed and replaced with a second batch of fresh food to ferment. The fermented food is washed and fibrous material removed. It is then dried in the sun for several hours to remove the volatile odours. It is then pounded into a paste. Grated coconut or coconut cream and sugar added and the mixture is wrapped in banana leaves and either baked or boiled.

During fermentation, carbon dioxide builds up in the pit, creating an anaerobic atmospHere. As a result of bacterial activity, the temperature rises much higher than the ambient temperature. The pH of the fruit within the pit decreases from 6. 7 to 3. 7 within about four weeks. Inoculation of the fruit in the pit with lactic acid bacteria greatly speeds up the process.

In the South Pacific pit fermentations are an ancient method of preserving starchy vegetables such as banana, plantain, breadfruit, cassava, taro, sweet potato, arrowroot and yams. The products undergo an acid fermentation, to produce a paste with good keeping qualities. It is usually pounded with a little sugar, coconut cream or fresh coconut and boiled or baked to make a type of pudding.

SINKI (PICKLED RADISH)

Sinki is a sour pickle prepared from radish tap roots. It is consumed traditionally in India, Nepal and parts of Bhutan, where it is used as a base for soup or eaten as a pickle. It is one of the most popular pickles in Nepal. Fresh radish roots are harvested, washed and wilted by sun-drying for one to two days. They are then shredded, re-washed and packed tightly into an earthenware or glass jar, which is sealed and left to ferment.

The optimum fermentation time is twelve days at 30°C. Sinki fermentation is initiated by L. fermentum and L. brevis, followed by L. plantarum. During fermentation the pH drops from 6. 7 to 3. 3. After fermentation, the radish substrate is sun- dried to a moisture level of about 21%. There is a second processing method involving fermentation in a clay lined pit for two to three months. For consumption, sinki is rinsed in water for two minutes, squeezed to remove the excess water and fried with salt, tomato, onion and green chilli. The fried mixture is then boiled in rice water and served hot as soup along with the main meal.

SUNKI

Sunki is a non-salted and fermented vegetable product prepared from the leaves of "Otaki-turnip" in Kiso district, Nagano prefecture, Japan. Sunki is eaten with rice and in miso soup. The Otaki-turnip is boiled, inoculated with "Zumi" (a wild small apple) dried Sunki from the previous year and allowed to ferment for one to two months. Sunki is produced under low temperature (in winter season). Micro-organisms involved include Lactobacillus plantarum, L. Brevis, Bacillus coagulans and Pediococcus pentosaceus.

KANJI

In Northern India and Pakistan carrots, especially a variety that is deep purple in colour, are fermented to make a traditional ready to serve drink known as kanji. Kanji is very popular and considered to have cooling and soothing properties and to be of high nutritional value.

After thorough washing the carrots are finely grated. Each kilogram of grated carrot is mixed with 7 litres of water, 200g of salt, 40g of crushed mustard seed and

8g of hot chilli powder. The mixture is then placed in a glazed earthenware vessel, which is almost entirely sealed, leaving only a tiny hole for gases released during fermentation to escape. The mixture is then allowed to ferment for seven to ten days. The type of fermentation that takes place is known as a lactic fermentation, which must be carried out in the absence of air.

Lactic acid bacteria produce lactic acid which reduces the pH (i.e. increases the acidity) to a level that prevents the growth of food poisoning organisms. The final product is slightly acidic in taste and has an attractive purple-red colour. After fermentation the drink is strained through fine muslin and has to be consumed within 3 or 4 days after which it goes bad. Each kg of grated carrot yields just over 7 litres of kanji.

FERMENTED TEA LEAVES

In South East Asia, tea leaves (Camellia sinensis) are fermented to make a sour-tasting snack. In Myanmar the product is called leppet-so, in Thailand it is known as miang.

ALKALINE BACTERIAL PRODUCTS

Product Description

Kawal is a strong smelling Sudanese, protein-rich food prepared by fermenting the leaves of a wild African legume, Cassia obtusifolia and is usually cooked in stews and soups. It is used as a meat replacer or a meat extender. Its protein is of high quality, rich in sulpHur amino acids which are usually obtained from either fish or meat.

Raw Material Preparation

The Sickle Pod plant (Cassia obtusifolia) is a wild legume that grows in Sudan. The leaves are collected late in the rainy season when the plant is fully grown. All the stems, pods and flowers are removed. The leaves are not washed, since it is thought that natural micro-organisms on the leaves are important for the correct fermentation.

Process and Principles of Preservation

The leaves of the leguminous plant are pounded into paste without releasing the juice. The paste is placed in an earthenware jar and covered with sorghum leaves. The whole jar is sealed with mud and buried in the ground up to the neck in a cool place. Every three days the contents are mixed by hand.

The fermentation takes about fourteen days. The fermentation is extremely complex. The main micro-organisms are Bacillus subtilis and Propionibacterium spp. Lactic acid bacteria including Lactobacillus plantarum; yeasts including Candida krusei and Saccharomyces spp and moulds including Rhizopus spp are also involved. After about fourteen days, the strongly smelling black fermented paste is made into small balls and sun-dried for five days.

Flow Diagram

Select leaves ↓	Remove stems and flowers
Grind leaves ↓	Pound the leaves into a paste in a mortar and pestle
Place in jar ↓	With sorghum leaves
Cover jar ↓	
Bury jar ↓	Bury up to the neck
Mix ↓	Every three days
Roll into balls ↓	
Sun dry ↓	Three to five days

OMBOLO WA KOBA

In Zaire cassava leaves are fermented to produce ombolo wa koba which is traditionally eaten with boiled cassava and plantain bananas. Cassava leaves are allowed to wilt and turn black. This takes about three to four days. The cassava leaves are then chopped up and placed in a pot of boiling water for about one hour. During this processing stage, a water soluble extract of ash is produced by placing the ash of burnt dried banana skins and palm tree flowers in a strainer and pouring water through it.

The extract is then added to the boiled cassava leaves. The extract is alkaline and neutralises the cyanhydric acid liberated when the leaves are chopped up. Salt and dried fish or meat is also added. After allowing the cassava leaf mixture to cool a little, acid palm oil is then added. This reacts with the excess alkali and neutralises it. The product is now ready to be eaten (Menea and Bishosha).

PRODUCTS OF MIXED FERMENTATIONS

Most traditional fermented food products are made by a complex interaction of different micro-organisms.

Vinegars

Vinegar is the product of a mixed fermentation of yeast followed by acetic acid bacteria. Vinegar, literally translated as sour wine, is one of the oldest products of fermentation used by man. It is the acetic acid produced by the fermentation of alcohol (ethanol) which gives the characteristic flavour and aroma to vinegar.

It can be made from almost any fermentable carbohydrate source, for example fruits, vegetables, syrups and wine. The basic requirement for vinegar production is a raw material that will undergo an alcoholic fermentation. Apples, pears, grapes, honey, syrups, cereals, hydrolysed starches, beer and wine are all ideal substrates for the production of vinegar. To produce a high quality product it is essential that the raw material is mature, clean and in good condition.

Indigenous vinegars can be made quite simply by the spontaneous fermentation of a fruit or alcohol. All that is necessary is an alcoholic substrate, strains of acetic-acid forming bacteria (acetobacter) and oxygen to enable the oxidation of alcohol. However, this process is very slow and vinegars produced by this method tend to be of inferior quality. Controlled fermentation conditions produce a more acceptable product. A wide range of raw materials can be made into vinegar.

COCONUT WATER VINEGAR

Raw Material Preparation

Coconut water is a waste product, which is produced in appreciable quantities in the Philippines, Sri Lanka Thailand and other countries. Its conversion into vinegar therefore presents an attractive option for decreasing wastage and producing a valuable product.

Processing

Coconut water is a good base for vinegar, but its sugar content is too low (only about 1%). Sugar needs to be added to bring the level of sugar up to 15%. After the addition of sugar, the coconut juice is allowed to ferment for about seven days, during which time the sugar is converted to alcohol. An alternative method is to pasteurise the coconut water and sugar mixture and add yeast. After this initial fermentation, strong vinegar (10% v/v) is added to stimulate the growth of acetic acid bacteria and discourage further yeast fermentation. The acetic acid fermentation takes approximately one month, yielding a vinegar with approximately 6% acetic acid. The fermentation will take less time than this if a generator is used. After fermentation, the vinegar must be stored in anaerobic conditions to prevent spoilage by the oxidation of acetic acid. Clarification can be achieved by stirring with a well beaten egg white, heating until the egg white coagulates and filtering (Anon).

PINEAPPLE PEEL VINEGAR

Product Description

This product enables the utilisation of pineapple peels, which are usually discarded during the processing or consumption of the fruit. The product has a distinct, very light pineapple flavour and has the same uses as any commercial vinegar.

Raw Material Preparation

The peels should be from very well washed ripe pineapples (damaged, rotten or infected fruits should not be used as a source of peels). Use only the peels, not the leaves or stems. The water used should be potable water, boiled if necessary. All the equipment should be well cleaned, as well as the bottles, which should also be steam-sterilised before use.

Processing

The peels should be cut into thin strips and put into clay or pewter pots. Aluminium or iron pots should not be used.

Sugar and clean water are added. Each pot is then inoculated and covered with a clean cotton cloth, held around the pot with an adhesive tape, to prevent contamination by insects or dust. The inoculated pineapple is fermented at room temperature (about 20°-22°C) for about eight days. The acidity should be checked daily. The water level should be maintained during this period. The product should be increasingly acid and by the eighth day it should have the required concentration of 4 per cent acetic acid in vinegar. If higher acidity is desired the product is left to ferment for another one or two days.

The development of acidity should be checked by tasting the product during fermentation. The residual bacteria removed may be reused as a residue inoculum two or three times more.

The traditional process may be improved by a two-stage fermentation in which alcohol is first formed by yeast (Saccharomyces cerevisiae) and the 'must' is then inoculated with acetic acid bacteria (Acetobacter pasteurianus). In outline, the process involves liquidising the peels and diluting with water (water: pulp is 4:1), adjusting the pH to 4.0 using sodium bicarbonate and adding yeast nutrient (ammonium pHospHate) at 0.14g per litre.

A starter culture is added at 2.7g per litre and the fermentation allowed to take place at 25°C for two days. The 'must' is then filtered and inoculated with acetic acid bacteria and allowed to ferment for eleven days with aeration of the 'must'. Other parts of the process are similar. Additional equipment includes a pH meter, refractometer, liquidiser, fermentation locks and equipment for preparing the starter cultures.

Flow Diagram

Pineapple ↓	Well washed in clean water
Peel the fruit ↓	Take care not to damage hands
Cut the peel ↓	Cut into thin strips and put into clay or pewter pots
Mix with sugar ↓	Sugar is dissolved in clean water
Ferment ↓	Each pot is then inoculated and covered with a clean cotton cloth
Filter ↓	Strain through a cheese cloth
Package	

PALM WINE VINEGAR

Palm wine vinegar is a produced across West Africa. It is a vinegar containing about 4% acetic acid, produced from the oxidation of palm wine. It is mainly consumed by people in urban areas as a salad dressing and meat tenderiser, although it also has medicinal uses and is valued in certain rituals. Palm wine is fermented using the same process as for grape wine vinegar the oxidation of alcohol to acetic acid. The spontaneous process takes about four days. The optimum fermentation temperature is 30°C.

COCONUT TODDY VINEGAR

Coconut toddy vinegar is produced throughout South Asia particularly Sri Lanka. It is a clear liquid with a strong acetic acid flavour and a hint of coconut flavour. The fresh toddy is strained, prior to allowing yeast fermentation to occur naturally for 48 to 72 hours. The yeast cells and debris are then removed by progressive sedimentation.

After two to four weeks of settling the fermented toddy is placed in barrels. The alcohol is then converted into acetic acid by acetic acid bacteria which are naturally present. The process can be hastened by adding vinegar as a starter. The fermented toddy is converted into vinegar in about three months. Ageing for six months, results in a pleasantly flavoured final product.

QUICK PROCESS PICKLES

Quick process pickles are easy to make but do not really constitute a fermented food product. For this technique, vegetables are soaked in a low salt solution for a few hours. They are then drained and placed in a container. The container is filled with a hot vinegar and spice mixture or a hot oil and spice mixture. There are hundreds of different recipes utilising locally available fruit and vegetables.

COCOA PRODUCTS

COCOA POWDER

Product Description

A fine brown powder with the characteristic taste of cocoa. It is a major ingredient in the confectionery and bakery industries. The product has a short shelf life. "Drinking chocolate" is a mixture of cocoa powder and sugar.

Raw Material Preparation

Cocoa beans are the seeds of the cocoa plant (Theobroma cacao). Cocoa pods are cut from the cocoa tree. The pods are cut and the beans removed. Only fully ripe and undamaged beans should be selected. It is important that the beans are processed quickly.

Processing

It was formerly believed that cocoa beans were fermented to remove the adhering pulp. However a good flavour in the final cocoa or chocolate is dependent on good fermentation. Fermentation is carried out in a variety of ways but all depend on heaping a quantity of fresh beans with their pulp and allowing micro-organisms to produce heat. The majority of beans are fermented in heaps although better results are obtained using boxes, which result in a more even fermentation.

Fermentation lasts from five to six days. During the first day the adhering pulp is liquified and drains away with the temperature rising steadily. The initial alcoholic fermentation gives way to acetification. This and other chemical changes cause the temperature to rise in excess of 50°C. The beans die. It was thought in the past that death was mainly due to increasing temperature. It is now known that acetic acid at a concentration of 1 percent in the bean is the cause of death and that it is only enhanced by heat, lactic acid and ethanol.

The pH value of the cotyledon drops from 6. 45 to 4. 5 over 120 hours and that during the same period the acetic acid content increased from 0 to 1.36 percent, while the lactic acid content increased from 0.005 to 0.12 percent. When the bean dies maceration of the tissue takes place, allowing enzymes and substrate to mix freely. The possible substrates for enzymes are carbohydrates, lipids, pHenolics and amino acids. In addition it is known that the bacteria can metabolise alcohols and organic acids of various kinds. The changing chemical picture is complex. Possible major substrates for micro-organisms are carbohydrates, lipids, pHenolics and amino acids. Unlike some flavours and aromas, that of chocolate is not attributable to a single compound.

During fermentation the external appearance of the beans changes. At first they are pinkish with a covering of white mucilage. Gradually the colour darkens and the mucilage disappears. The beans on the surface are always darker than those deeper in the heap or box, indicating that the colour change is oxidative. As the beans are mixed, their colour becomes a more uniform orange-brown and they are only slightly sticky. At this stage they are ready for drying.

The beans need to be dried to a moisture content of less than 7.5%. The beans are dried by either being spread out in the sun in layers a few centimetres thick or in artificial dryers. There are numerous types of dryers but it is important that any smoky products of combustion do not come in contact with the beans otherwise taints will appear in the final product.

The beans are cleaned to remove the extraneous matter. Cocoa beans consist of an outer skin that needs to be removed and inner "nib". The shell is sometimes removed before roasting and sometimes after roasting.

The cake from the mill is ground in a hammer mill to produce the cocoa powder.

Flow Diagram

Cocoa beans ↓	
Sort ↓	Select only mature beans
Ferment ↓	In heaps or boxes
Clean ↓	Remove extraneous material
Dry ↓	In the sun or in artificial dryers
Roast ↓	120 to 150°C
Grind ↓	
Remove cocoa butter ↓	
Grind cake ↓	
Pack	

For cocoa powder roasting temperatures of 120 to 15°C are used. There are many designs of roasters: both batch and continuous systems. The operation is controlled so that the cocoa is heated to the required temperature without burning the shell or the cotyledon. The heat is applied evenly over a long period of up to 90 minutes to produce even roasting. The bean must not be contaminated with any combustion products from the fuel used and provision must be made for the escape of any volatile acids, water vapour and decomposition products of the bean. After roasting the beans are cooled quickly to prevent scorching.

The roasted nibs are ground into a powder in a plate mill. The resulting powder is sieved through fine silk, nylon or wire mesh. To produce cocoa powder, some of the cocoa butter needs to be removed. With low fat cocoa powder, more than 90% of the cocoa butter is removed. With medium fat cocoa powder, more than 78% of the cocoa butter has been removed. Finally high fat cocoa powder has less than 78% of the cocoa butter removed. Extrusion, expeller, or screw presses are used in the cocoa industry to remove the cocoa

Packaging and Storage

Cocoa powder is hygroscopic (picks up moisture from the air) and should be protected, especially in humid climates. Lidded tins or sealed polythene bags should be used.

CHOCOLATE

Preparation of Raw Materials

Cocoa beans are the seeds of the cocoa plant (Theobroma cacao). Cocoa pods are put from the trees, and the beans are removed from the pods. Only fully ripe and undamaged beans should be selected. It is important that the beans are processed quickly.

Processing

For cocoa butter production the roasting temperatures are 100 C to 104 C There are many designs of roasters: both batch and continuous systems. The operation is controlled so that: the cocoa is heated to the required temperature without burning the shell or the cotyledon The heat is applied evenly over a long period of up to 90 minutes to produce even roasting; the nib must not be contaminated with any combustion products from the fuel used and provision must be made for the escape of any volatile acids, water vapour and decomposition products of the nib. After roasting the beans are cooled quickly to prevent scorching

Roasting will have already loosened the shell. The beans are then lightly crushed with the object of preserving large pieces of shell and nib and avoiding the creation of small particles and dust. The cocoa bean without its shell is known as a "cocoa nib". The valuable part of the cocoa bean is the nib, the outer shell being a waste material of little value.

Alkalization is a treatment that is sometimes used before and sometimes after grinding to modify the colour and flavour of the product. This was developed in the Netherlands in the last century and is sometimes known as "Dutching". This involves soaking the nib or the cocoa mass in potassium or sodium carbonate. By varying the ratio of alkali to nib, a wide range of colours of cocoa powder can be produced. Complete nib penetration may take an hour. After alkalization the cocoa needs to be dried slowly.

The cocoa nib is ground into "cocoa liquor" (also known as "unsweetened chocolate" or "cocoa mass"). The grinding process generates heat and the dry granular consistency of the nib is turned into a liquid as the high amount of fat contained in the nib melts. There are various pre-treatments to develop the flavour of the cocoa mass with and without reaction solutions. These include the "Luwa thin-layer evaporator", "Petzomat thin-layer process", "Cocovap process", "Lehman KFA process" and "Carle-Montanari process".

Extrusion, expeller, or screw presses are used in the cocoa industry for the production of cocoa butter from whole beans, and mixtures of fine nib dusts, small nibs, and immature beans. Research in the Kerala Agricultural University has led to develop a suitable pressure device capable of separating cocoa butter from ground cocoa mass ideally suitable for small scale manufacturers. In Peru a simple screw press is used to extract cocoa butter from beans. The crude cocoa butter is filtered through cloth and allowed to solidify. To produce plain chocolate, cocoa mass is mixed with sugar and sufficient cocoa butter to enable the chocolate to be moulded. The ratio of mass to sugar varies according to the national taste. The mixture is ground to such a degree that the chocolate is smooth to the palate. At one time this was done by a lengthy process in melengeurs heavy granite rollers in a revolving granite bed but nowadays grinding is done in a series of rolls.

The chocolate is then "Conched". This may last for several hours. The chocolate is heated, this helps to drive off volatile acids, thereby reducing acidity when present in the raw bean, and the process finishes the development of flavour and makes the chocolate homogeneous. Similar processes are involved in the manufacture of milk chocolate. The milk is added in various ways either in powder form to the mixture of mass, sugar and cocoa butter, or by condensing first with sugar, adding the mass and drying this mixture under vacuum. The product is called 'crumb' and this is ground and conched in a similar manner to plain chocolate. After conching the chocolate has to be tempered before it is used for moulding or for enrobing confectionery centres. Tempering involves cooling and reaching the right pHysical state for rapid setting, after moulding or enrobing.

COFFEE

Raw Material Preparation

Coffee beans are harvested from two plants Coffea arabica and Coffea canepHora variety robusta. Only ripe berries should be used in coffee production.

Berries can be placed in water so that immature berries which float can be identified and discarded.

Processing

Dry processing is the simpler of the two processing methods and is popular in Brazil for the processing of Robusta coffee and in Sri Lanka for processing Arabica coffee. The coffee cherries are dried immediately after harvest by sun drying on a clean dry floor or on mats. The dried berry is then hulled to remove the pericarp. This can be done by hand using a pestle and mortar or in a mechanical huller. The mechanical hullers usually consist of a steel screw, the pitch of which increases as it approaches the outlet so removing the pericarp. The hulled coffee is cleaned by winnowing.

Wet processing involves squeezing the berry in a pulping machine or pounding in a pestle and mortar to remove the outer fleshy material (mesocarp and exocarp) and leave the bean covered in mucilage. This mucilage is removed by fermentation. Fermentation involves placing the beans in plastic buckets or tanks and allowing them to sit, until the mucilage is broken down. Natural enzymes in the mucilage and yeasts and bacteria in the environment work together to break down the mucilage. The coffee should be stirred occasionally and every so often a handful of beans should be tested by washing in water. If the mucilage can be washed off and the beans feel gritty rather than slippery, the beans are ready.

There is much debate about the fermentation of coffee beans. Some researchers feel that the mucilage breakdown is caused by enzymatic breakdown. For instance Wellman has stated that enymatic fermentation starts immediately the beans have been squeezed from the fresh berries. If these 'pulped' beans are piled up or put in a container and protected from any bacterial or other contamination the fermentation will progress.

After a number of hours the enzymes of the pulp will have acted on the torn tissues, gorged with starches, sugars and pectins, in such a manner that, without any microbial intervention, the remaining pulp will be easily detached from the beans and washed off in water. However most investigators acknowledge the necessity for the presence of micro-organisms for the depectinisation of the beans.

The following micro-organisms have been isolated: Leuconostoc mesenteroides; Lactobacillus plantarum; Lactobacillus brevis, Streptococcus faecalis, Aerobacter (Enterobacter) and Escherichia, pectinolytic species of Bacillus, Saccharomyces marscianus, S. bayanus and a Flavobacterium sp., Erwinia dissolvens, Fusarium spp, Aspergillus spp and Penicillium.

The beans should then be washed immediately as 'off' flavours develop quickly. To prevent cracking the coffee beans should be dried slowly to 10% moisture content (wet basis). Drying should take place immediately after to prevent 'off' flavours developing. The same drying methods can be used for this as for the dry processed coffee. After drying the coffee should be rested for 8 hours in a well ventilated place. The thin parchment around the coffee is removed either by hand, in a pestle and mortar or in a small huller. The hulled coffee is cleaned by winnowing.

The final flavour of the coffee is heavily dependent on how the beans are roasted. Roasting is a time temperature dependent process. The roasting temperature needs to be about 200 C. The degree of roast is usually assessed visually. One method is to watch the thin white line between the two sides of the bean, when this starts to go brown the coffee is ready. As preferences vary considerably from region to region, a lot of research will need to be done to find the locally acceptable degree of roast. Coffee beans can be roasted in a saucepan as long as they are continually stirred. A small improvement is made by roasting the coffee in sand, as this provides a more even heat. A roaster will produce a higher quality product.

Grinding is a means of adding value to a product. However, it is fraught with difficulties. It is easy to make an assessment of an intact bean, while a ground product presents some difficulty. The fear of adulteration and the use of low quality produce is justified. Because of this there is a great deal of market resistance to ground coffee. This market resistance can only be overcome by consistently producing a good product. There are basically two types of grinders - manual grinders and motorised grinders.

Flow Diagram

Coffee cherry ↓	
Sort the cherries ↓	Select only mature undamaged beans
Pulp ↓	Skin and pulp removed
Ferment ↓	To remove mucilage
Dry the beans ↓	Sun drying or artificial drying
Roast ↓	
Grind ↓	
Package	

Packaging and Storage

Roasted beans can be stored in sacks. Milled beans need to be packaged quickly to prevent the loss of volatile flavour components. The packaging material should be airtight. Polythene is not suitable as it is a low barrier to loss of aroma.

OTHER MIXED FERMENTATION PRODUCTS

Vanilla

Vanilla is produced in Madagascar, Indonesia and various South Pacific islands. It is a dark brown pod about 20 cm in length. Vanilla is produced by fermenting the pods of the orchids of the genus vanilla. The pods are first sun dried for 24 to 36 hours and then blanched in hot water (65^{0}C) for two to three minutes. The pods are then fermented in boxes and dried again.

Tabasco

Tabasco sauce is made in Mexico and Guatemala. The chilli pods are harvested, ground into a paste and placed in a container with salt. The hot and fiery sauce develops.

Tea

In the production of tea, there is a process referred to as fermentation. However microbial activity is not involved in the so-called 'fermentation' of tea. The chemical changes are effected by enzymes alone. Fermentation rooms are used where moisture and temperature can be controlled. During fermentation even further darkening of the leaf occurs and the typical aroma develops. By subjective judgement of the aroma's intensity the period necessary for completion is gauged.

CHAPTER 10

Food and Physical Activity

Staying at-or getting to-a healthy weight helps us in several ways. Remember: Not only does it help us get to that "better me," but research shows it plays an important role in reducing the risk of several types of chronic diseases. If you are overweight or obese, you are not alone. Many of us are. In fact, two-thirds of adults are overweight or obese. That is why many of us need to eat fewer calories, be more physically active, and make wiser food choices.

LOSING WEIGHT, GAINING LIFE

Lifestyle changes in diet and pHysical activity are the healthiest choices for weight loss. To lose weight, many of us need a reduction in 500 calories or more per day from food and drink. In addition, increasing pHysical activity is also important. Remember: To lose weight, calories in must be less than calories out. When it comes to weight control, it is calories that count-not the proportions of fat, carbohydrates, and protein in the diet. Diets that provide very low or very high amounts of protein, carbohydrates, or fat are likely not providing enough of some nutrients-that is why they are not advisable for long-term use. Although these kinds of weight-loss diets have been shown to result in weight loss, maintaining weight loss ultimately depends on a change in your lifestyle. A healthy lifestyle is about more than just your weight it is also about feeling better and reducing your risk of chronic diseases.

ENERGY BALANCE

To maintain your body weight, the energy that you expend through your daily activities (for example, breathing, sleeping, and moving around) and additional pHysical activities has to equal the energy (food and drink) that you consume as calories. If these two things-energy use and energy consumption-are equal, then you are in "energy balance."

If you want to lose weight, you will have to either increase the amount of energy expended through additional pHysical activities or decrease the calories that you consume, or both. Be careful to still get all the nutrients that you need if you decide to eat fewer calories to tip the energy balance. The recommendations provide the right proportions of fat, carbohydrates, and protein for good health. Remember: It is always important to eat foods that are high in nutrients for the number of calories they contain, such as fruits, vegetables, whole grains, and fat-free or low-fat milk or equivalent milk products.

We know it's difficult to lose weight; it is better not to gain it in the first place. Since many of us tend to gain weight slowly as we age, small decreases in calorie intake can help avoid that slow weight gain, especially when accompanied by increased pHysical activity. Remember: 3,500 calories equals 1 pound. For most adults, a reduction of 50 to 100 calories per day from foods and beverages may prevent gradual weight gain.

CALORIE-LOWERING STRATEGIES

The healthiest way to reduce calorie intake is to reduce intake of added sugars, saturated and trans fats, and alcohol, which all provide calories but few or no "good for you" nutrients. You already know about added sugars and unhealthy fats.

Alcoholic beverages supply calories but few essential nutrients. Calories are provided for serving volume of beer, wine, and distilled spirits. Higher alcohol content (higher percent alcohol or higher proof), and mixing alcohol with other beverages, such as sugar-sweetened soft drinks or tonic water, fruit juice, or cream, increases the number of calories in the beverage. Remember: You need to reduce your food and drink intake by 3,500 calories or burn 3,500 calories without increasing food intake to lose a pound. If you have one light beer each day, that is 100 extra calories! A cocktail made with a sugar-sweetened mixer could be twice that number of calories or more!

Beverage	Approximate Calories (per 1 fluid oz)	Example Serving Volume	Approximate Total Calories
Beer (regular)	12	12 oz	144
Beer (light)	9	12 oz	108
White wine	20	5 oz	100
Red wine	21	5 oz	105
Sweet dessert wine	47	3 oz	141
80-proof distilled spirits (gin, rum, vodka, and whiskey)	64	1.5 oz	96

The same goes for those of us who drink sugar-sweetened beverages without any alcohol. Many of them have at least 150 calories in 1 serving-usually one 12-fluid ounce can. But often, these beverages come in larger bottles, and we drink the entire bottle-easily 300 to 500 unnecessary calories. If you have one 20-ounce sugar-sweetened beverage each day, in 2 weeks, it can add up to over 3,500 excess calories-or a weight gain of 1 pound.

Another way to reduce calorie intake is to eat foods that are low in calories for the amount of food eaten. Examples of these foods are many kinds of vegetables and fruits and some soups. If a soup is made with cream, it will not count toward this strategy, but to double the impact, a soup made with low-sodium broth and packed with vegetables can be low in calories, packed with nutrients, and filling too! Just watch out for. the salt content. Remember: It is always important to eat foods that are high in nutrients for the number of calories they contain, such as fruits, vegetables, whole grains, and fat-free or low-fat milk or equivalent milk products.

One more way to reduce calorie intake is to evaluate the portion size of foods. We already know, portion size is the amount of food eaten at one sitting-and, in general, portion size has increased significantly over the past two decades. For example, the muffin you ate at work 5 years ago may have had 300 calories, but today's muffins are typically larger and may contain up to 500 calories. You may have not noticed the increase-but it could be 200 extra calories! And, science shows us that controlling portion size helps limit calorie intake, particularly when eating calorie-dense foods (foods that are high in calories for a given measure of food).

Therefore, it is essential that we understand how portion size-the amount you eat-compares to a recommended standard amount of food (that is, a serving) from each food group at a specific calorie level. One technique that is useful in looking for ways to cut calories from foods is to write down what we eat. Do it for 1 day. This means writing down everything you eat, along with an estimate of the amount you eat-this time be realistic. If you put a pat of butter on your toast, write down both the toast and a teaspoon of butter-that is, if your pat was a teaspoon! How do you know how much of a food you are eating? Take a minute to look in your cabinet. Find your measuring spoons and cups. Next, find something familiar that you can use to eyeball measurements that you will remember.

BE MORE ACTIVE

Eating fewer calories, of course, is just one side of the equation. Calorie output needs to be in balance with calorie intake to maintain body weight. That means the calories "burned" during pHysical activity (in addition to activities of daily living) need to be the same as the calories taken in as food. Here are a few recommendations about how much pHysical activity we should be doing and how "hard" we should be performing. But since we have been talking energy balance, we can't wrap up the discussion without talking about pHysical activity a little bit.

So, here we go! For health purposes, all adults should engage in at least 30 minutes of moderate- intensity pHysical activity on most days of the week. This recommended amount of activity is in addition to light-intensity routine activities of daily living and pHysical activity of less than 10 minutes. However, to prevent weight gain, adults may need up to 60 minutes of moderate to vigorous-intensity activity on most days of the week while at the same time not eating more calories than required. To sustain weight loss, adults may need as much as 60 to 90 minutes of moderately intense pHysical activity per day while not eating too much.

MAKING PHYSICAL ACTIVITY PART OF A HEALTHIER YOU

Most of us know we need to be physically active to be healthy. It's not new information, but it leaves us with many questions and many opinions. There's plenty of information about pHysical activity, but sorting through it and figuring out what to do can be challenging. Instead of answering questions, all that information only seems to generate more: "What exactly does being physically active mean? Is this pHysical activity or 'exercise? ' How much do we need versus what can we do to get by? Do we need to do it all at once? Is there an easy way to fit it into our day, because life is pretty hectic already? " So many questions... plus, almost every day, a new exercise product is introduced promising a quick fix-to whip us into shape requiring only the slightest amount of effort on our part. If only that were true, we tell ourselves....

Here's what we can agree on-we'll give it to you straight: The basic, scientifically grounded information on pHysical activity. Then, we'll begin to figure out how to balance this with your day and your lifestyle.

THE BASICS

Science tells us that when it comes down to our overall health, adults, regardless of age, need to do two key types of pHysical activities: We are going to introduce a couple of terms that you may be less familiar with in this context-moderately intense and vigorously intense. By understanding how much effort we need to exert, we can begin to choose what kind of activity fits our available time, life, and needs. Here are some tips and examples to help identify whether pHysical activity is moderately or vigorously intense.

CARDIO-WHAT'S YOUR INTENSITY?

Moderate: While performing the pHysical activity, if your heart is beating noticeably faster-it's probably moderately intense: We need to do this level of activity for at least 30 minutes, most days of the week.

Examples include:

- Walking briskly (a 15-minute mile)
- Light yard work (raking/bagging leaves or using a lawn mower)
- Light snow shoveling
- Actively playing with children
- Biking at a casual pace.

Vigorous: If you are breathing hard and fast and your heart rate is increased substantially during pHysical activity, it's probably vigorously intense.

Examples include:

- Jogging/running
- Swimming laps
- Rollerblading/ inline skating at a brisk pace
- Cross-country skiing
- Most competitive sports (football, basketball)
- Jumping rope.

You don't have to do 30 minutes all at once...

To meet the goal of 30 minutes a day of moderately intense pHysical activity, you don't have to do all 30 minutes at once. Scientific evidence shows you get the same health benefits from breaking 30 minutes up into three 10-minute or two 15-minute intervals throughout the day, if you prefer. Daily activities like climbing several flights of stairs or parking farther away from store entrances are a good start. But if you don't do that activity for at least 10 minutes at a time, it doesn't help you meet the recommendation for 30 minutes of moderately intense pHysical activity. In addition, for most people, greater health benefits can be obtained by engaging in pHysical activity of more vigorous intensity or of longer duration.

WHAT IS STRENGTH TRAINING AND WHY DO IT?

Moving on to the second type of pHysical activity-strength training. This consists of resistance exercises, which means your muscles work or hold against an applied force or weight. These exercises increase the strength of our muscles, help maintain the integrity of our bones, and may improve our balance, coordination, arid mobility. Strength training can also change the appearance of our bodies with better muscle definition.

Usually, resistance exercises are done using weights or workout bands, but there are other ways to achieve the same effect. If you like going to the gym, include resistance exercises in your routine there. Strength training or resistance exercises can be done anywhere. An example of a goal to work toward could be to do 8 to 12 repetitions of 6 to 8 strength-training exercises twice a week.

Examples include:

- Push-ups
- Pull-ups
- Biceps curls
- Sit-ups
- Carrying full laundry baskets
- Rowing a boat
- Strength training in aerobics class.

Both strength training and cardio activity are important. In particular, strength training helps develop and maintain healthy bones, and develop and tone muscles.

WHAT'S IN IT FOR ME

Part of it comes down to: Feeling better! Looking better! Not so bad, right? In the big picture, it also comes down to good pHysical health. If that's not enough, there's also that sense of well-being you get from regular pHysical activity-a constructive way to deal with the demands of the day, release stress, and just feel better about yourself. Many people say that exercising regularly helps them have more energy, sleep better, and simply, enjoy taking time to do something good for themselves.

With most everything, extra work really does pay off! Physical activity is no exception, and the more active you are, the more you benefit. For example, you can further reduce your risk for many chronic diseases, including cardiovascular disease, type 2 diabetes, colon and breast cancers, and osteoporosis, by doing more than the minimum 30 minutes of moderate-intensity pHysical activity on most days of the week. Incorporating up to 60 minutes of cardio activity may also help you to prevent unhealthy weight gain or to manage your weight, if that is your goal.

Some of us suffer from the yo-yo factor-weight on, weight off, weight on, weight off. Sure, we can lose the weight, but how do we keep it off for good? That may take at least 60 to 90 minutes of daily moderate-intensity pHysical activity. Sounds like a lot-no kidding! But it's the truth based on the data of people who have successfully lost weight (at least 30 pounds) and kept it off for at least a year. Keep your perspective and start small-do your pHysical activity in 10- minute moderately intense increments and build up. Eventually, you will become your own success story.

Different intensities and types of exercise offer different benefits. Cardio or aerobic activities exercise your heart and increase your ability to be physically active for a longer period of time. This type of endurance makes it easier to carry out harder tasks for longer periods of time-whether it's keeping up with your kids or grandchildren, or playing basketball with your co-workers. Strength training or resistance exercises also contribute to muscular endurance. Strength training is especially beneficial as we get older.

As we age, we tend to lose bone and muscle mass, making it difficult to carry out everyday activities: getting in and out of a chair, carrying groceries or laundry, or just walking. Together, cardio and strength training work your whole body. Vigorous pHysical activity (for example, jogging or other aerobic exercises) provides greater health benefits for pHysical fitness than does moderate pHysical activity and burns more calories per unit of time. Aside from all the health benefits, what a bonus that it also seems to make us feel better about ourselves.

Now to the biggest challenge: How do we fit this into our life? Our busy, already-pressed-for-time, on-a-budget life? A lot of people have shared their thoughts with us. Here's some of their feedback on what works.

Buddy System

Some days it's hard to talk yourself into an activity. Working with others who are going through the same thing can be motivating, especially when you promised that you would meet for a walk in the park, or a tennis match, or signed up to take a yoga class together. You don't want to let your buddy down. In the process, you end up not letting yourself down either. Buddies can be co-workers, spouses, neighbors, or even faraway friends that you stay in touch with via e-mail and provide encouragement. Heck, your walking buddy can even be your pet!

The Great Outdoors

Opportunities for pHysical activity may be closer than you think. Take advantage of public parks and pools. There are millions of acres to explore-walk, hike, swim, kayak, canoe, and bike. Also consider being a volunteer. Whether it's leading a hike or cleaning a trail or playground, you are making a difference in your life and others'.

Enjoy What You Do

If aerobicizing in a room full of people isn't your thing, why do it? There are hundreds of activities to choose from. Find something you like and chances are you will stick with it. Maybe you were a swimmer when you were younger, but haven't thought of it since high school. Many local park and recreation facilities, or area schools, have open or lap swims.

Some people enjoy walking around a nearby school track, or if you prefer indoors, walk at your local mall. What about hiking at a local, state, or national park; playing in a soccer, volleyball, or softball league; or taking a yoga, Pilates, or tai-chi class a couple days a week? Don't limit yourself. Don't get discouraged. Pick a few activities to try out, rotate them, and slowly you will figure out what works best for you. Trying something new can be fun and give you more confidence to pursue other activities.

HEALTHIER OLDER ADULTS

If you're an older adult or perhaps playing a vital role in taking care of aging parents or grandparents, there are some nutrition and pHysical activity considerations to keep in mind. Getting older doesn't mean that our quality of life or desire to be our best is any different. In fact, we usually become more aware of our health as we age. Healthful habits can help older adults enjoy daily activities, stay mobile, and be independent. Anytime is a good time to start healthy habits, no matter how old we are. So, if you are a little older... you can still be healthier. Eating a balanced diet of nutrient packed foods applies to all of us, but for older adults, a healthy eating plan may require a little more planning. If you have health problems or take medication regularly, it may be important to check with your health care provider for advice about changing your diet or pHysical activity level.

FIBRE, MORE IMPORTANT THAN EVER

We've talked about how a healthy diet includes fibre-rich foods, such as fruits, vegetables, and whole grains, that offer many health benefits including protection against heart disease. Another benefit is that fibre promotes regularity. Constipation may affect older adults for many reasons-from taking certain medications to drinking less fluid.

How much fibre do you need? The recommended dietary fibre intake is 14 grams per 1,000 calories consumed. So, the more calories you eat, the more fibre your body needs. Good sources of dietary fibre include: cooked dry beans and ready-to-eat bran cereal or shredded wheat; pears and berries; dried prunes, figs, and dates; and cooked green peas, Brussels sprouts, sweet potatoes, and spinach. For a 2,000 calorie diet, you will need 2% cups, of vegetables (a source of fibre and other nutrients) each day. Consuming at least 3 or more ounces of whole grains can reduce the risk of several chronic diseases and may help with weight maintenance.

FATS AND YOUR HEART

Many of us, especially if we are older, have been told to eat less fat. Fat can impact the health of our heart and arteries in positive and negative ways, depending on the type of fat. All the more reason to stay away from saturated fats, trans fats, and cholesterol.

Eating too much saturated and trans fats, the type of fats that are solid at room temperature, may increase the risk of heart disease. Saturated fats can be found in animal-based products such as milk and milk products, butter, meat, and poultry. And eating too much cholesterol, a fatty substance found only in animal-based products, may also increase the risk of heart disease. It's important to eat less than 10 percent of your calories from saturated fats.

For example, if you aim to eat 2,000 calories per day, your daily allowance of saturated fat would be less than 200 calories or 20 grams-which equals 100% Daily Value (DV) for saturated fat. And, remember, this is a limit, not a goal, meaning you do not need to achieve your DV for saturated fat! Furthermore, you should keep trans fats (often found in cakes, cookies, crackers, pies, and breads) as low as possible, and eat less than 300 milligrams per day (mg/day) of cholesterol.

Maybe you are someone who has an elevated LDL (bad) cholesterol level. Definitely, you should follow your health care provider's advice. Those of us with elevated cholesterol may be advised to decrease our calories from saturated fat to less than 7 percent of total calories-which is about 16 grams or about 80% DV and less than 200 mg/day of cholesterol. It's critical to find out what's right for a Healthier You.

Now, a few words about making wise fat choices: an immediate change you can make is to eat mono-unsaturated and polyunsaturated fats found in fish, nuts, and vegetable oils to reduce saturated fat calories in your diet.

In fact, to help reduce the risk of heart disease, some evidence suggests eating approximately 2 servings of fish per week (a total of about 8 ounces) for people who have already had a heart attack. It may reduce their risk of death from cardiovascular disease.

RELATION BETWEEN SODIUM AND POTASSIUM

Nearly all of us eat too much salt (sodium). As a matter of fact, on average, the more salt we eat, the higher our blood pressure-and most of the salt we eat comes from processed foods, not necessarily from the salt shaker. Surprised? Eating less salt is an important way to reduce the risk of high blood pressure, which may in turn reduce the risk of heart disease, stroke, congestive heart failure, and kidney damage.

Other lifestyle changes may prevent or delay getting high blood pressure. These include eating more foods rich in potassium, losing excess weight, being more physically active, and eating an overall healthy diet.

A diet rich in potassium helps to counterbalance some of sodium's harmful effects on blood pressure. You may need to consult your health care provider for advice on how much sodium and potassium you should get, but in general, older adults should aim to consume no more than 1,500 mg/day (about 60% DV on the food label) of sodium, and meet the potassium recommendation of (4,700 mg/day) by eating potassium-rich food. When choosing packaged foods, check the sodium content on the Nutrition Facts label. Older adults should not exceed about 60% DV for sodium for the day.

KEEPING YOUR FOOD SAFE

Did you know that, every year, an estimated 76 million people in the United States become ill from food that contains harmful bacteria? Older adults are at higher risk. Perhaps, food borne illness has affected you, and you did not even recognize the common symptoms-an upset stomach, diarrhea, a fever, vomiting, abdominal cramps, and dehydration. It can also result in more severe illness, such as paralysis and meningitis, or even death.

It's important that older adults, people with weakened immune systems, and individuals with certain chronic illnesses pay extra attention and carefully follow food safety advice. Here are simple steps that you and your family can take to minimize the risk-four key words: clean, separate, cook, and chill. In addition, older adults need to avoid eating or drinking raw (unpasteurized) milk or any products made from unpasteurized milk, raw or partially cooked eggs or foods containing raw eggs, raw or undercooked meat and poultry, raw or undercooked fish or shellfish, unpasteurized juices, and raw sprouts.

BRINGING IT TOGETHER

We've talked a lot about the healthy eating plan: getting a variety of fruits, vegetables, whole grains, and fat-free or low-fat milk and equivalent milk products; including lean meats, poultry, fish, legumes (dry beans and peas), eggs, nuts, and seeds; and balancing calorie intake with calorie needs. Each major food group provides a variety of nutrients, so it's important to include all food groups in your daily eating plan.

For older adults, every day it's important to eat:

- Fibre-containing foods
- Vitamin B12-fortified foods, such as fortified cereals, or take the crystalline form of vitamin B12 supplements
- Vitamin D-fortified foods, such as fortified milk or orange juice, which are important for calcium absorption and can reduce the risk of bone loss.

In general, older adults should get enough foods that contain calcium, potassium, fibre, magnesium, and vitamins A,C,D, and E, without eating too many foods high in calories, saturated and trans fats, cholesterol, added sugars, and salt (sodium).

We've already talked about some adjustments to make to our diets. Now, let's quickly review how this fits into your daily calorie needs based on "My Personal Profile," or if you are caring for older adults, take a moment to figure out their range. For example, a 60-year-old, sedentary woman should aim for 1,600 calories per day, while a 60-year-old, sedentary male should aim for 2,000 calories per day. When increasing the amount of fruits, vegetables, and legumes you eat, be sure to eat them in place Of less nutritious foods, not in addition to them, if weight control is part of your goal. Next, let's work in some pHysical activity.

BE ACTIVE

Many older adults may feel that they are too tired to be physically active or that they have earned their rest. However, pHysical activity is a critical part of a healthy lifestyle, and for older adults, pHysical activity may take on even more meaning. Continuing to live independently-doing the things that you enjoy can be linked to being active. Increasing your heartbeat, strengthening your muscles, and increasing your flexibility contribute to pHysical fitness and the ability to do everyday activities like climbing the stairs, shopping for groceries, and visiting with family and friends.

Older adults may want to consult with their health care provider, if they have certain chronic diseases or are taking medications, before starting vigorous pHysical activity. Let's also clear up a misconception-that older adults should not participate in pHysical activity because of a risk of falls or injury. Actually, the opposite is true. Sedentary older adults have a higher risk of falls and regular pHysical activity may reduce their risk. Research also shows that regular pHysical activity can promote psychological well-being and can aid in reducing feelings of mild-to-moderate depression and anxiety.

On a day that you're feeling a bit tired, down, or stressed, consider taking a brisk walk around your neighborhood or at the mall. Start small, have a positive attitude, build up to more vigorous activities, and continue to enjoy all that life has to offer you! Let's get into the nitty gritty... for adults ages 65 and older, here's what science tells us to do:

- At a minimum, do moderately-intense cardio or aerobic activity for at least 30 minutes per day, most days of the week.
- At a minimum, do strength-training exercises, 2 days per week.

Moderate Activity

Intensity is relative to your level of fitness. For some older adults, moderately-intense pHysical activities include:

- Walking briskly
- Biking at a casual pace
- Dancing (ballroom, line dancing, jazz, or tap)
- Water aerobics
- Golfing without a cart
- Light gardening/yard work such as raking or pushing a power lawn mower
- Actively playing with children
- Doubles tennis.

Vigorous Activity

For some older adults, vigorously-intense pHysical activities include:

- Jogging
- Swimming laps
- Singles tennis
- Heavy yard work.

STRENGTH TRAINING

Strength-training exercises are resistance exercises that increase the strength of your muscles, help maintain the integrity of your bones, and may improve your balance, coordination, and mobility. Both strength training and cardio or aerobic activities are important. In particular, strength training helps develop and maintain a healthy skeleton and muscle mass. A few examples of strength-training exercises are:

- Digging in a garden
- Chopping wood
- Using a push lawn mower
- Bicep curls
- Yoga (some types are more strengthening than others)
- Leg lifts
- Squats.

BALANCE TRAINING

Some exercises improve your balance and strength at the same time. If you are at risk of falling, you should include balance exercises as part of your strength-training activities. Some examples include:

- Rising up and down on your toes while standing and holding a stable chair or countertop
- Walking a straight line heel to toe
- Side leg raises while standing and holding onto a chair
- Knee flexions (while standing and holding onto a chair, bend knee so your foot lifts behind you)
- Hip flexions (while standing and holding onto a chair, raise knee toward chest).

HEALTHIER CHILDREN

Many of us touch the lives of children in some way. We have an opportunity to give children the building blocks they need to lead a healthy lifestyle. As parents, doctors, educators, extended family, or friends-we are all in a position to set good examples. By teaching children the importance of good nutrition and regular pHysical activity early, they'll learn good habits to last a lifetime. Whether it's a family effort or simply taking advantage of the time spent with children, healthy choices start with all of us.

Since children are growing, sometimes it's hard to know their weight status-are they overweight or will they "grow" out of it? We know that maintaining a healthy weight throughout childhood and adolescence may reduce the risk of becoming an overweight or obese adult.

Always check with your child's health care provider about your child's rate of growth and development before starting him or her on any type of weight-gain or weight-loss diet. A health care provider can give you good nutrition and weight management approaches that take into consideration that your child is growing and developing. This is especially important if your child has a medical condition or is on medication.

FOR CHILDREN

The keys to healthy eating are variety, balance, and moderation. Just as you determined the amount of food from each food group that was right for you, children and adolescents have amounts that are right for them, too. You can estimate the amount of calories your child should be eating each day as you did for yourself. Here are some additional considerations to take when finding a healthy eating plan for your child.

Whole Grains

Everyone, including kids, should consume whole-grain products often, and at least half the grains they eat should be whole grains. Think whole-grain cereals and sandwiches made with whole wheat bread, but be sure to check the ingredient list for "whole wheat."

Calcium

Because calcium is important for growing bones, children 2 to 8 years should consume 2 cups per day of fat-free or low-fat milk or equivalent milk products. Children 9 years and older should consume 3 cups per day of fat-free or low-fat milk or equivalent milk products.

Fats and Salt

It's important for us to know how much total fat and the types of fat we eat (for example, saturated and trans fats) and the same holds true for children. When kids reach 2 years, it is time to start watching the types of fat and how much salt they eat. Children who eat high saturated fats or salt (sodium) diets can be at risk of high blood cholesterol and high blood pressure just like adults. That's why total fat intake should be kept between 30 and 35 percent of calories for children ages 2 to 3 years, and between 25 and 35 percent of calories for children and adolescents ages 4 to 18. Most fats should come from sources of polyunsaturated and monounsaturated fatty acids, such as fish, nuts, and vegetable oils. Following the eating plans, at the calorie needs for your children, will help them get the nutrients they need to grow up healthy.

Cavity (caries) Prevention

Consider how many sugary snacks (think cavities!) your children are eating. Sugary snacks contain calories, but few or no essential nutrients, and may increase the chance that your kids will get cavities. Some helpful tips to reduce cavities: regular brushing, fluoride toothpaste, and fluoridated water. Also replace sugary snacks with healthy snacks-fruits, vegetables, and whole grains can make a difference.

Food Safety

There are a couple of special considerations when it comes to children and food safety. Infants and young children should not eat or drink raw (unpasteurized) milk or any products made from unpasteurized milk, raw or partially cooked eggs or foods containing raw eggs, raw or undercooked meat and poultry, raw or undercooked fish or shellfish, unpasteurized juices, and raw sprouts.

A SMART START

We've all had to balance, at one time or another, what children want with what's best for them. Nobody wants to feel like they are denying a child certain types of food. Making smart choices doesn't have to come down to this. Here are some tips and strategies that might help reach a healthy balance:

- Try to keep track of children's meal/snack and pHysical activity patterns so you can encourage a healthful lifestyle.
- For younger children, the family setting is a good place to encourage them (along with the rest of the family) to eat a variety of differently colored vegetables arid fruits each day. You can start their day with 100% fruit or vegetable juice. Slice fruit on top of a whole-grain cereal. Serve a salad with lunch and an apple as an afternoon snack. Include vegetables with dinner.
- Remember: Fill your shopping cart with fruits, vegetables, nuts, and fat-free or low-fat milk and limit candy, soft drinks, chips, and cookies. Children will soon learn to make these types of smart food choices outside the home as well.
- Choose a variety of foods. No single food or food group supplies all the nutrients in the amounts that you need for good health. If you plan for pizza one night, include other food or food groups by adding a salad, serving fat-free or low-fat milk, and having fruit for dessert.

EAT SMART

Eating together is an ideal way for family and friends to spend time together and enjoy each other's company. Whether you're eating at home or eating out on the go, it's important to eat smart. Check off some of the tips you can easily put into action. Try one or two at a time. Don't bite off more than you can chew!

- Be consistent. Establish a family meal routine, and set times for breakfast, lunch, dinner, and snacks. Eat together whenever possible.
- Take charge of the foods children eat. When you serve a meal, a child can choose to eat it or not; but don't offer to substitute a less healthy alternative when the child refuses to eat what you've served.
- Limit children's access to the refrigerator and snack cupboards.
- Turn off the TV during meals, and limit children's snacking when watching TV. It is easy to lose track of how much they are eating, especially if the snack is a bag of pretzels instead of a small bowl.
- Serve a vegetable or fruit with every meal and at snack time.

- Reward children with praise and fun activities rather than food.
- Involve children in meal planning and food preparation. They are more likely to eat what they help to make.
- While shopping and cooking, teach children about the food groups and the importance of a healthy eating plan. Throughout the day, help them choose the types and amounts of foods they need from the different food groups.
- Teach children how to read food labels and use the quick guide to the percent Daily Value (% DV): 5% DV or less is low; 20% DV or more is high.
- Use low-fat cooking methods such as baking, roasting, and grilling; and when you use oil, select those such as olive or canola oils.
- Serve water and fat-free or low-fat milk, with and between meals. Children over 2 years should be encouraged to drink fat-free or low-fat milk. Whole milk is recommended for children from 1 to 2 years old. Check with your health care provider for feeding recommendations for younger children and those with special needs.
- Teach children how to make wise food choices away from home-at school cafeterias, restaurants, and vending machines.
- Teach children to pay attention to both the quality and quantity of their food choices. More food is not always better; they need to understand appropriate portionsizes.

Good nutrition and regular pHysical activity should be part of an overall healthy lifestyle. To grow healthfully, children must balance the calories they eat with what they use up being physically active.

PLAY TIME CAN BE PHYSICAL ACTIVITY TIME.

Moderate	Vigorous
Light yard work	Running
Walking	Swimming
Bicycling	Basketball

Play is an important part of growing and developing. It allows children to learn, explore, and be physically active. All of this is critical for children to help strengthen muscles, bones, and joints, and it gives them the opportunity to gain confidence while having fun! Children need to get at least 60 minutes of pHysical activity on most, if not all, days of the week. Playing hopscotch, tossing a ball back and forth, and dancing, in addition to organized sports, are some good ways for your child to be active and learn important life skills along the way. Some examples of activities listed by intensity of pHysical activity. Your children can always step up the intensity by working harder!

- Be a physically active role model and have fun with children. Since adults need 30 minutes of daily pHysical activity, play together! Then, make sure the child gets in another 30 minutes for a total of 60. Another 30 minutes is probably good for you, too.

- Walk with children at every available opportunity-if possible to school or to the store on errands. Take a walk with family or friends after dinner instead of watching TV or playing computer games.
- Plan active weekends. Include biking, hiking, skating, walking, or playing ball. Take a trip to the park, swimming pool, or ice skating rink. If you work on weekends, arrange a physically active play date for your child.
- Offer to join children in their favourite pHysical activities, or enroll children in a group exercise programme.
- Include children in active chores such as dog walking, house cleaning, car washing, and yard work.
- Limit physically inactive behaviour such as TV watching and computer time.
- Avoid using TV as a child-sitter or pacifier. Offer active alternatives to screen time-jumping rope, playing hide-and-seek, or running an errand. Children love it when you are active with them and share in what they do.
- Keep TVs out of children's rooms.
- Give your children gifts that encourage pHysical activity-active games or sporting equipment.
- Talk with your schools about ways to incorporate non-competitive pHysical activity during the day.

An easy and fun way to keep children active and eating right is to create a weekly calendar of healthy lifestyle activities. Let everyone choose a weekly activity and take charge of it.

Index